K

274 | DON | 8.3.04

20 054 043 49

AF759235

Controlling disease due to helminth infections

edited by

D.W.T. Crompton
WHO Collaborating Centre for Soil-transmitted Helminthiases
Institute of Biomedical and Life Sciences
University of Glasgow, Glasgow, Scotland

A. Montresor
Strategy Development and Monitoring for Parasitic Diseases and Vector Control
Control, Prevention and Eradication
World Health Organization, Geneva, Switzerland

M.C. Nesheim
Division of Nutritional Sciences
Cornell University, Ithaca, NY, USA

L. Savioli
Strategy Development and Monitoring for Parasitic Diseases and Vector Control
Control, Prevention and Eradication
World Health Organization, Geneva, Switzerland

WORLD HEALTH ORGANIZATION
GENEVA
2003

WHO Library Cataloguing-in-Publication data

Controlling disease due to helminth infections / edited by
D. W. T. Crompton . . . [et al.].

1.Helminthiasis—prevention and control 2.Helminthiasis—drug therapy 3.Anthelmintics—pharmacology 4.National health programs 5.Epidemiologic studies 6.Evaluation studies I.Crompton, David W. T. II.Controlling Disease due to Helminth Infections (2000 : Bali, Indonesia)

ISBN 92 4 156239 0 (NLM classification: WC 800)

NORTHUMBRIA UNIVERSITY
LIBRARY
ITEM No. CLASS No.
2005404349 616.96 CON

© **World Health Organization 2003**

All rights reserved. Publications of the World Health Organization can be obtained from Marketing and Dissemination, World Health Organization, 20 Avenue Appia, 1211 Geneva 27, Switzerland (tel: +41 22 791 2476; fax: +41 22 791 4857; e-mail: bookorders@who.int). Requests for permission to reproduce or translate WHO publications—whether for sale or for noncommercial distribution—should be addressed to Publications, at the above address (fax: +41 22 791 4806; e-mail: permissions@who.int).

The designations employed and the presentation of the material in this publication do not imply the expression of any opinion whatsoever on the part of the World Health Organization concerning the legal status of any country, territory, city or area or of its authorities, or concerning the delimitation of its frontiers or boundaries. Dotted lines on maps represent approximate border lines for which there may not yet be full agreement.

The mention of specific companies or of certain manufacturers' products does not imply that they are endorsed or recommended by the World Health Organization in preference to others of a similar nature that are not mentioned. Errors and omissions excepted, the names of proprietary products are distinguished by initial capital letters.

The World Health Organization does not warrant that the information contained in this publication is complete and correct and shall not be liable for any damages incurred as a result of its use.

The named authors and editors alone are responsible for the views expressed in this publication.

Designed by minimum graphics New Zealand
Typeset by SNP Best-set Typesetter Ltd., Hong Kong
Printed by Interpint, Malta

When Carlo Urbani took it upon himself to investigate the new and deadly disease we now know as SARS (severe acute respiratory syndrome) and so alert the world to this emerging danger, he did so without the slightest regard for his own health and safety. He died of SARS in Bangkok on 29 March 2003 aged 46, at the pinnacle of his career – first and foremost as a physician but also as an authority on the prevention and control of parasitic disease.

Carlo Urbani now stands shoulder to shoulder with Edward Jenner, Ignaz Semmelweis, Louis Pasteur, Robert Koch, Rudolph Virchow, Matthew Lukwiya and other giants of public health whose efforts have relieved us from the burden of infectious disease.

Carlo contributed to this volume, which is dedicated to his memory with affection and admiration.

Contents

Preface

This book originated from the presentations and discussions that took place at a conference held in Bali, Indonesia, in February 2000. The conference was convened and administered by the Government of Indonesia and the World Health Organization, with technical assistance from Cornell University, Ithaca, NY, USA, and the University of Glasgow, Scotland. The main purpose of the conference was to review activities for the control of disease due to soil-transmitted helminth infections in Indonesia and neighbouring countries. This theme was particularly timely and apt since, in November 1999, the Cabinet of the World Health Organization had concluded that there should be a scaling up of worm control to encompass soil-transmitted helminthiasis and schistosomiasis. Earlier impetus for such a conference had already come from the G8 summits held in 1997 and 1998, at which the then Prime Minister of Japan, Mr Hashimoto, had called on member governments to increase aid and assistance for the control of parasitic diseases in developing countries.

The book is not intended as a comprehensive guide to planning and implementing helminth control programmes, nor is it a simple compilation of conference proceedings. Some of the articles were invited from experts who were unable to attend the conference. However, we hope that the contents will encourage fellow health professionals to reappraise the public health significance of helminth-induced disease in their regions and note that deworming is a feasible and cost-effective option, even when resources are in very short supply. The contents seeks to draw attention to: the public health significance of a selection of helminth infections; current ideas about approaches to the control of helminth-induced disease; information about anthelminthic drugs and their use; and recent experiences of control programmes in several countries.

Although there was little discussion in Bali about dracunculiasis, lymphatic filariasis, onchocerciasis, or schistosomiasis, these topics are included in the book. We hope to urge health managers to consider the integration of control measures for soil-transmitted helminth infections into existing health care programmes; the new initiative to eliminate lymphatic filariasis, for example, offers opportunities to include measures for the control of disease caused by soil-transmitted helminths.

Many friends and colleagues have earned our sincere thanks for all the efforts that led to the success of the Bali conference. We thank Fasli Jalal and his staff from the Ministry of Planning in Indonesia, the WHO Secretariat in Indonesia, and Muriel Gramiccia at WHO in Geneva for preparing for the conference and for making arrangements to bring delegates to Bali and support them there. We thank all the contributors to the book and colleagues who reviewed articles before publication. Patricia Peters and Desho Sahonta in Glasgow deserve special thanks for their invaluable secretarial assistance with the book's production. We are equally grateful to Sarah Ballance, who undertook to see the volume through all stages of its production. Finally, we thank the Government of Japan for generous financial support.

D.W.T. Crompton
A. Montresor
M.C. Nesheim
L. Savioli

Opening address

The purpose of this important, international conference is to promote and achieve further significant progress in controlling the diseases caused by endoparasitic helminth infections. We have gathered to share experience of political, planning, and practical measures to extend advocacy and to make recommendations with the aim of relieving needy people from the burden of helminthiases.

In 1605, my fellow countryman, Miguel de Cervantes, published his monumental contribution to world literature. He introduced us to Don Quixote, an errant, foolhardy, and ill equipped knight who galloped in futile fashion to tilt at a windmill. Until quite recently, our profession might have accused itself of tilting at windmills. We did not have the knowledge and the tools to tackle helminthiasis. This situation has changed, as we shall hear during this conference. There have, however, been some notable landmarks in the challenge against helminthiases. I will mention a few and draw attention to some important lessons, which need to be learned and remembered. We must not ignore our history.

The Rockefeller Sanitary Commission was founded in 1909, almost a century ago, with the aim of eradicating hookworm disease in the southern states of the USA and beyond. Critics have claimed that the Commission failed; hookworm is still with us. But please note, the workers in the Commission sought to eradicate—perhaps they would have been wiser to have aimed at control—disease, not infection. If you scrutinize the Commission's reports and publications you will find their actions relied on the best possible epidemiological evidence, they monitored the results of interventions, and the whole enterprise was based on private financial donations. Sadly, the key tools, modern anthelminthic drugs, were not available to the workers of the Rockefeller Sanitary Commission.

Following establishment of the People's Republic of China in 1949, the Government and population, despite extremely limited resources, settled down to deal with parasitic diseases. Of particular significance has been the striking success in reducing the public health impact of *Schistosoma japonicum.* In this case, political will, with the support and cooperation of the people, made a major contribution to bringing relief from an exceptionally nasty parasitic disease.

In 1968, a famous meeting in Tunis was the catalyst for what has become the Onchocerciasis Control Programme in West Africa. Apart from bringing relief from the ravages of this disease, the programme has been sustained; public health workers have proved, often under difficult circumstances, that the momentum and the initial enthusiasm of a control programme can be kept going. Recently, we have witnessed massive progress in the drive to eliminate dracunculiasis. Advance planning is now taking place to initiate the global elimination of lymphatic filariasis by the year 2020. The World Health Organization, NGOs, governments, and pharmaceutical industries are collaborating closely to ensure success. We have demonstrated that entirely different types of institution can focus on a common aim to work together. In practice, public health issues like parasite control will not succeed without multifaceted collaboration.

Overall, however, we must look to and learn from our friends and colleagues in Japan if we wish to win victories in the war against worms. At the end of the Second World War, the Government of Japan, the academic, scientific, and medical communities, together with the people, settled down to rid the country of ascariasis. Japanese agencies and organizations went out to develop parasite control programmes across Asia and beyond. Fundamental and operational research into parasite control has been supported. A wealth of knowledge and experience is available for approaching the control of disease caused by parasitic helminths, thanks to the efforts of our Japanese colleagues. In 1998, the Hashimoto Initiative was launched, following a presentation on the need to control the burden of parasitic diseases worldwide. This issue was placed before the leaders of the world's wealthiest nations meeting at the G8 summit in Birmingham, England. The World Health Organization, on behalf of billions of poor people in developing countries, thanks Japan for its past, present, and future efforts to control the global burden of parasitic diseases.

The World Health Organization is scaling up its role in the effort to control disease due to helminth infections. The Organization sees this conference in Bali as another landmark in the process. On behalf of the Director-General of the World Health Organization, Dr Gro Harlem Brundtland, myself and all of the delegates, I must sincerely thank the Government of Indonesia, the Governor of Bali, the organizers, and those responsible for financial support for making this conference a reality. I look forward to the proceedings, the recommendations, and their implementation.

Dr Maria Neira
Director, Communicable Diseases Control, Prevention and Eradication
World Health Organization
February 2000

The Bali Declaration

25 February 2000

A conference of international experts has agreed that deworming programmes relieve the burden of disease caused by infection with parasitic worms and bring about major improvement in personal and public health.

Benefits accrue from deworming, in childhood growth, development, and cognition, in adult productivity, and in the course and outcome of pregnancy.

The Bali conference declares that the World Health Organization, as a matter of urgency, should call on the governments of the developed countries to contribute to relieving poor people worldwide of this unnecessary burden of disease.

Similarly, governments of developing countries should include parasite control programmes as a matter of high priority on their national agendas.

List of participants and contributors

Controlling Disease due to Helminth Infections

Bali, Indonesia

20–24 February 2000

S.A.N. Abidin, Department of Parasitology, Faculty of Medicine, University of Indonesia, Jakarta, Indonesia

Z. Agus, School of Medicine, Andala University, Padang, Indonesia

M. Albonico, Fondazione Ivo de Carneri, Turin, Italy

T.M.S. Atukorala, Department of Parasitology, Faculty of Medicine, University of Colombo, Colombo, Sri Lanka

Phyu Phyu Aye, School Health Division, Department of Health, Ministry of Health, Yangon, Myanmar

G.P. Bordignon, World Food Programme, West Africa Regional Bureau, Dakar, Senegal

Le Dinh Cong, National Institute of Malariology, Parasitology and Entomology, Tu Liem, Hanoi, Viet Nam

L. Chitsulo, Parasitic Diseases and Vector Control, Communicable Diseases Control, Prevention and Eradication, World Health Organization, Geneva, Switzerland

H.M. Chwaya, Preventive Services, Ministry of Health, Zanzibar, United Republic of Tanzania

D.W.T. Crompton, WHO Collaborating Centre for Soil-Transmitted Helminthiases, Institute of Biomedical and Life Sciences, University of Glasgow, Glasgow, Scotland

Nguyen Thi Dang, Department of Helminthology, National Institute of Malariology, Parasitology and Entomology, Tu Liem, Hanoi, Viet Nam

L.C. David, Internal Planning Service, Department of Health, Santa Cruz, Manila, Philippines

Nguyen Van De, Department of Helminthology, National Institute of Malariology, Parasitology and Entomology, Tu Liem, Hanoi, Viet Nam

N.R. de Silva, Department of Parasitology, Faculty of Medicine, University of Kelaniya, Ragama, Sri Lanka

L. Drake, Partnership for Child Development, Imperial College School of Medicine, London, England

Nguyen Thi Duc, Department of Helminthology, National Institute of Malariology, Parasitology and Entomology, Tu Liem, Hanoi, Viet Nam

D. Engels, Parasitic Diseases and Vector Control, Communicable Diseases Control, Prevention and Eradication, World Health Organization, Geneva, Switzerland

M.M. Gyi, School Health Division, Department of Health, Ministry of Health, Yangon, Myanmar

T. Gyorkos, Division of Clinical Epidemiology, Montreal General Hospital, Montreal, QC, Canada

V. Hadju, Pusat Studi Gizi dan Pangan, Gedung Pusat Kegiatan Penelitian, Kampus Unhas Tamalanrea, Makassar, Indonesia

T. Hara, Japan Association of Parasitic Control, Tokyo, Japan

B. Hatz, Swiss Tropical Institute, Basle, Switzerland

Sri Hendratno, Partnership for Child Development, Department of Infectious Disease Epidemiology, Imperial College School of Medicine, London, England

I. Hettiarachchi, Plantation, Health and Social Welfare Trust, Colombo, Sri Lanka

Nguyen Thi Viet Hoa, Department of Helminthology, National Institute of Malariology, Parasitology and Entomology, Tu Liem, Hanoi, Viet Nam

M.E. Hodges, c/o D.W.T. Crompton, WHO Collaborating Centre for Soil-Transmitted Helminthiases, Institute of Biomedical and Life Sciences, University of Glasgow, Glasgow, Scotland

J. Horton, TDR/PRD, World Health Organization, Geneva, Switzerland

S. Hoyer, World Health Organization, Phnom Penh, Cambodia

K. Ichimori, Division for Combating Communicable Diseases, WHO Regional Office for the Western Pacific, Manila, Philippines

M.M. Ismail, Department of Parasitology, Faculty of Medicine, University of Colombo, Colombo, Sri Lanka

F. Jalal, Center for School Feeding Project, National Development Planning Agency (Bappenas), Jakarta, Indonesia

T. Khamkeo, Hygiene and Preventive Department, Ministry of Public Health, Vientiane, Lao People's Democratic Republic

Myo Khin, Department of Medical Research, Ministry of Health, Yangon, Myanmar

Win Win Khine, Department of Medical Research, Ministry of Health, Yangon, Myanmar

Hoang Thi Kim, Department of Helminthology, National Institute of Malariology, Parasitology and Entomology, Tu Liem, Hanoi, Viet Nam

J.D. Kvalsvig, Child, Youth and Family Development, Human Sciences Research Council, Durban, South Africa

Y.C.F. Lumampao, Soil-transmitted Helminth Control Programme, Center for Infectious Diseases, Department of Health, Santa Cruz, Manila, Philippines

M. Mahaswiati, Central Laboratory, Yayasan Kusuma Buana, Jakarta, Indonesia

Sri S. Margono, Department of Parasitology, Faculty of Medicine, University of Indonesia, Jakarta, Indonesia

M. Mogensen, Partnership for Child Development, Department of Infectious Disease Epidemiology, Imperial College School of Medicine, London, England

A. Montresor, Parasitic Diseases and Vector Control, Communicable Diseases Control, Prevention and Eradication, World Health Organization, Geneva, Switzerland

T. de S. Naotunne, Department of Parasitology, Faculty of Medicine, University of Colombo, Colombo, Sri Lanka

M. Neira, Communicable Diseases Control, Prevention and Eradication, World Health Organization, Geneva, Switzerland

M.C. Nesheim, Division of Nutritional Sciences, Cornell University, Ithaca, NY, USA

P. Odermatt, Médecins Sans Frontières, Phnom Penh, Cambodia

Sri Oemijati, Parasitology Unit, Faculty of Medicine, University of Indonesia, Jakarta, Indonesia

K. Palmer, WHO Regional Office for the Western Pacific, Manila, Philippines

K. Pholsena, Hygiene and Preventive Department, Ministry of Public Health, Vientiane, Lao People's Democratic Republic

D. Ratnawulan, Central Laboratory, Yayasan Kusuma Buana, Jakarta, Indonesia

Han-Jong Rim, Korea Association of Health, Seoul, Republic of Korea

A. Sasongko, Yayasan Kusuma Buana, Jakarta, Indonesia

Satoto, Partnership for Child Development, Department of Infectious Disease Epidemiology, Imperial College School of Medicine, London, England

L. Savioli, Parasitic Diseases and Vector Control, Communicable Diseases Control, Prevention and Eradication, World Health Organization, Geneva, Switzerland

Thaw Saw, Department of Medical Research, Ministry of Health, Yangon, Myanmar

D.R. Shakya, School Feeding Programme Team, World Food Programme, Kathmandu, Nepal

N. Sham-Laye, Environmental Health Section, Ministry of Health, Victoria, Seychelles

San Shwe, Department of Medical Research, Ministry of Health, Yangon, Myanmar

M. Sinoun, CNM National Malaria Centre, Ministry of Health, Phnom Penh, Cambodia

D. Socheat, Médecins Sans Frontières, Phnom Penh, Cambodia

H. Subagyo, Partnership for Child Development, Department of Infectious Disease Epidemiology, Imperial College School of Medicine, London, England

T. Takeuchi, Department of Tropical Medicine and Parasitology, School of Medicine, Keio University, Tokyo, Japan

Hoang Van Tan, Department of Helminthology, National Institute of Malariology, Parasitology and Entomology, Tu Liem, Hanoi, Viet Nam

A.R. Thaha, Pusat Studi Gizi dan Pangan, Gedung Pusat Kegiatan Penelitian, Kampus Unhas Tamalanrea, Makassar, Indonesia

Tin Tin Than, Department of Medical Research, Ministry of Health, Yangon, Myanmar

Thein Hlang, Trauma Services, Parkview Hospital, Fort Wayne, In, USA

Nguyen Van Tien, Department of Helminthology, National Institute of Malariology, Parasitology and Entomology, Tu Liem, Hanoi, Viet Nam

H. Torlesse, Helen Keller International, Gulshan, Dhaka, Bangladesh

Aung Tun, School Health Division, Department of Health, Ministry of Health, Yangon, Myanmar

Pham Thi Tuyet, Department of Helminthology, National Institute of Malariology, Parasitology and Entomology, Tu Liem, Hanoi, Viet Nam

C. Urbani, Vector-borne and other Parasitic Diseases, World Health Organization, Hanoi, Viet Nam

Hkwar Nyo Zin, School Health Division, Department of Health, Ministry of Health, Yangon, Myanmar

Zen Rafilluddin, Partnership for Child Development, Department of Infectious Disease Epidemiology, Imperial College School of Medicine, London, England

Observers

Y.T. Chang, Shin Poong Pharmaceutical Company Ltd, Seoul, Republic of Korea

K. Kiikuni, Sasakawa Memorial Health Foundation, Tokyo, Japan

Unable to attend

D.A.P. Bundy, Human Development Network, World Bank, Washington, DC, USA

P. Hadidjaja, Department of Parasitology, Faculty of Medicine, University of Indonesia, Jakarta, Indonesia (*represented by*: S.A.N. Abidin)

A. Hall, Partnership for Child Development, Imperial College School Of Medicine, London, England

T.K. Kasai, Infection Control Division, Japanese Ministry of Health and Welfare, Tokyo, Japan

F. Lubis, Yayasan Kusuma Buana, Jakarta, Indonesia

C. Nokes, Partnership for Child Development, Imperial College School Of Medicine, London, England

E. Pollitt, Department of Pediatrics, TB-139 Program in International Nutrition, University of California at Davis, Davis, CA, USA

T. Sato, Division of Tuberculosis and Infectious Diseases, Ministry of Health and Welfare, Tokyo, Japan

R.J. Stoltzfus, Center for Human Nutrition, Department of International Health, School of Hygiene and Public Health, Johns Hopkins University, Baltimore, MD, USA

PART I

Public health significance

Public health significance

The United Nations estimates that 182 million preschool children—33% of those living in developing countries—are stunted when their height-for-age is compared with the norms for well-nourished children living in good environments (ACC/SCN, 2000). Underweight and stunted children are known to be at greater risk of dying during childhood (Pelletier, 1994) and may not achieve their full potential in education and physical performance (Martorell & Scrimshaw, 1995).

A child's height and weight are determined in part by the complex relationship between the child, the child's diet, and the environment. Tomkins & Watson (1989) described a malnutrition-infection complex—infection reduces food intake, leading to reduce growth rate, diarrhoea, and reduced immune function.

Malnutrition and parasitic infection occur concurrently where poverty ensures the persistence of poor housing, low levels of education, poor health services, inadequate sanitation, and lack of clean water (Crompton & Nesheim, 1982). It is now accepted that parasitic disease is a major contributor to the etiology of the malnutrition-infection complex (Crompton, 1993). Research has established that helminth infections are accompanied by reduced growth rates during childhood and by impaired nutrient utilization; hookworm infections affect iron status, often to the point where anaemia develops.

Progress has been made in raising the priority accorded to helminth control, and it is important to continue to explore public health aspects of helminthiases. The contributors to Part I of this book review additional evidence of the significance of disease due to helminth infections. Much of the growth faltering in children occurs between the ages of 6 months and 2 years—in other words, at the time when several helminth infections begin to be established and the immune system is challenged to respond. The more can be learned about child-helminth interactions, the greater will be the success of efforts to reduce the burden of disease due to helminths.

References

ACC/SCN (2000) *Fourth report on the world nutrition situation: nutrition throughout the life cycle.* Geneva, United Nations Administrative Committee on Coordination, Sub-Committee on Nutrition.

Crompton DWT, ed. (1993) Human nutrition and parasitic infection. *Parasitology*, 107:S1–S190.

Crompton DWT, Nesheim MC (1982) Nutritional science and parasitology: a case for collaboration. *Bioscience*, 32:677–680.

Martorell R, Scrimshaw NS (1995). The effects of improved nutrition in early childhood: the INCAP follow-up study. *Journal of Nutrition*, 125:4S.

Pelletier DL (1994) The potentiating effects of malnutrition on child mortality: epidemiologic evidence and policy implications. *Nutrition Reviews*, 52:409–416.

Tomkins AM, Watson F (1989) Malnutrition and infection: a review. London, London School of Hygiene and Tropical Medicine (ACC/SCN State-of-the-Art Series, Nutrition Policy Paper No. 5).

Important human helminthiasis in Indonesia

Sri S Margono[1]
Department of Parasitology
Faculty of Medicine, University of Indonesia
Jakarta, Indonesia

Introduction

Many species of helminth have been reported as causing infections in humans in Indonesia; however, only some of these species are highly prevalent and widely distributed. Of the nematode infections, both soil-transmitted helminthiasis and lymphatic filariasis are public health problems in the country. Among the trematodes, *Schistosoma japonicum* is present in areas of Sulawesi and *Fasciolopsis buski* in Kalimantan, while the cestodes *Taenia saginata* and *T. solium* are found in several provinces.

Soil-transmitted helminths

The tropical climate of Indonesia is highly favourable to the persistence of soil-transmitted helminths. The most important species infecting people in this country are *Ascaris lumbricoides*, *Trichuris trichiura*, and *Necator americanus*. *Ancylostoma duodenale* is rare compared with *N. americanus*; when it is found, it is generally in mixed infections with *N. americanus*. This phenomenon may be related to the temperature needed for eggs to develop into larvae in the soil: the optimum temperature for *A. duodenale* eggs is 23–25°C, but in most areas of Indonesia the temperature is higher throughout the year.

Surveys over the period 1970–1980 found prevalences of more than 70% for *Ascaris* and *Trichuris* infections (Carney et al., 1974; Cross et al., 1976; Margono et al., 1979). However, prevalences were very low in Nusa Tenggara Timur: 10% for *Ascaris* and 1% for *Trichuris* infections (Carney et al., 1975), possibly because of the dry climate in this province. The prevalences of *Ascaris* and *Trichuris* in a group of children under 5 years of age in Joglo, Jakarta, were respectively 73.2% and 60.9% (Margono & Ismid, 1987); in an urban area of Kramat, Central Indonesia, corresponding figures for 90 children of the same age were 66.67% and 61.12% (Ismid, 1996). In North Jakarta, 762 stool samples from five primary schools were examined for intestinal helminths. The prevalence of *Ascaris* was

[1] Professor Dr Sri S Margono, Department of Parasitology, Faculty of Medicine, University of Indonesia, Jl Salemba Raya No. 6, Jakarta 10430, Indonesia
Tel: +62 21 310 2135/+62 21 829 1870, Fax: +62 21 391 4607/+21 829 1870,
E-mail: margonofk@yahoo.com

59.96% and of *Trichuris* 79.64%; however, values for eggs per gram (epg) of faeces—an indicator of intensity of infection—were respectively 2719 and 232 (Ismid, Margono & Abidin, 1996).

In Mataram, Lombok, high prevalences of *Ascaris* were found in two schools (78.5% and 72.6%); for *Trichuris* infection the values were 63.95% and 60.0%. The epg values in the two schools were respectively 5192 and 7170 for *Ascaris*, and 572 and 2247 for *Trichuris* infections before treatment (Ismid et al., 1999). In East Lombok too, high prevalences were detected in two primary schools: 73.45% and 96.57% for *Ascaris*, and 69.03% and 79.43% for *Trichuris* infections. The epg values in these schools were respectively 2102 and 3883 for *Ascaris* infections, and 205 and 436 for *Trichuris* (Margono et al., 2000, unpublished data).

Low prevalences and epg values were reported from five urban schools in Jakarta. Examination of 549 stool samples from primary-school children (grades 3, 4, and 5) showed prevalences of 4.3% and 10.7%, respectively, for *Ascaris* and *Trichuris* infections. The epg values were also low—236, 387, and 87 for *Ascaris* infections in grades 3, 4, and 5, and 31, 55, and 15, respectively, for *Trichuris* infections (Margono et al., 2000).

At school 1 in Talang Dabok and at school 2 in Sungai Rengit, South Sumatra, prevalences of *Ascaris* infections were 40.3% and 58.9% respectively. Most of the *Ascaris* infections were very light (89%) or light (11%) at school 1; infections for school 2 were categorized as very light (81%), light (14%), moderate (1%), and heavy (4%). For *Trichuris* infections, prevalences were 41.0% and 75.9%, respectively, for school 1 and school 2. At school 1, 47% of the infections were very light, 42% light, and 11% moderate, and at school 2, 50% were very light, 48% light, and 2% heavy before treatment. Lower prevalences of *Ascaris* and *Trichuris* were found among the children at school 1, whose parents were more educated and had better personal hygiene behaviour and better environmental sanitation—but were financially worse off—than the parents of children at school 2 (Margono et al., 1998).

Among schoolchildren in two municipal areas of West Sumatra, 58.6% and 58.6% were infected with *Ascaris*, whereas 73.7% and 48.7% were infected with *Trichuris*. For two rural areas, prevalences were 6.6% and 55.2% for *Ascaris* infections, and 36.8% and 42.9% for *Trichuris* infections (Agus et al., 1999[1]). In urban slum areas of South Sulawesi, Hadju[2] (1999) found 92% and 98% of the schoolchildren were infected with *Ascaris* and *Trichuris*, respectively; corresponding figures for poor rural areas were respectively 66% and 57%.

[1] Agus Z et al. (1999) *The effectiveness of mass treatment of soil-transmitted helminthiases among school children in West Sumatra*. Report of meeting with Bappenas (Badan Perancang Pembangunan Nasional—the National Planning Board).

[2] Hadju V (1999) *Dampak intervensi pemulihan kecacingan di wilayah PMT-AS Sulawesi Selatan*. Report for Bappenas (Badan Perancang Pembangunan Nasional—the National Planning Board).

Findings for prevalence of *Ascaris* and *Trichuris* infections in various areas of Indonesia, and for epg values, are summarized in Tables 1 and 2.

In 1987, Margono & Ismid reported on children living in a peripheral area of Jakarta, which has since become more urban and has a higher population density; 17.4% of the children under 5 years of age were infected with hookworm. On a coffee plantation in East Java, hookworm prevalence was 36% among children of this age, and the youngest infected child was 8 months old (Tantular, 1984, personal communication). In general, prevalences of hookworm infection among schoolchildren were low in Jakarta: Margono (1986) reported prevalences of 2.4–10.6% in a slum area, while Margono, Ismid & Rukmono (1989) reported a figure of 0.6% for a suburban area. However, higher prevalences, 6.9–29.4%, were

Table 1. Prevalences of *Ascaris* and *Trichuris* infections in primary-school children in several areas of Indonesia

Area	Year	Prevalence (%)		Reference
		Ascaris	*Trichuris*	
North Jakarta	1996	59.96	79.64	Ismid et al., 1996
South Sumatra	1998	40.3	41.0	Margono et al., 1998
		58.9	75.9	
West Lombok	1999	78.5	64.0	Ismid et al., 1999
		72.6	60.0	
East Lombok	1999	73.45	69.03	(unpublished data)
		96.57	79.43	
South Sulawesi	1999	92	98	Hadju, 1999[a]
		66	57	

[a] Hadju V (1999) *Dampak intervensi pemulihan kecacingan di wilayah PMT-AS Sulawesi Selatan.* Report for Bappenas (Badan Perancang Pembangunan Nasional—the National Planning Board).

Table 2. Intensity of *Ascaris* and *Trichuris* infections in primary-school children in several areas of Indonesia

Area	Year	Eggs per gram (epg) of faeces		Reference
		Ascaris	*Trichuris*	
North Jakarta	1996	2719	232	Ismid et al., 1996
West Lombok	1999	5192	572	Ismid et al., 1999
		7170	2247	
East Lombok	1999	2102	205	(unpublished data)
		3883	436	
East Jakarta	2000	236	31	Margono et al. (in press)
		387	55	
		87	15	

reported for a rural area of Jakarta (Margono, 1986). In a rural area of South Sumatra, high prevalences of hookworm infection were also found among schoolchildren. Of 139 stool samples from school 1 and 83 from school 2, 38.8% and 39.7% respectively were positive for hookworm larvae. The infections were very light, however, with epg values of 32 and 37 (Margono et al., 1998). Low prevalences of hookworm infections were reported among primary-school children in two municipal and two rural areas of West Sumatra. In the municipal areas the prevalences were 2.5% and 5.2%, and in the district areas 5.3% and 8.6% (Agus et al., 1999). Hadju (1999) reported that 1.4% and 0.5%, respectively, of schoolchildren in urban slums and poor rural areas of South Sulawesi were infected with hookworm.

Observations on hookworm infection in various areas of Indonesia are summarized in Tables 3 and 4.

Control programmes covering several areas of Indonesia have been in effect for several years. Conducted by the Government through local health units, these programmes distribute anthelminthics and provide health education to school-

Table 3. Prevalence of hookworm infections in children under 5 years of age in East Java and Jakarta, Indonesia

Area	Year	Prevalence (%)	Reference
East Java	1984	36	Tantular, 1984 (personal communication)
Jakarta	1987	17.4	Margono & Ismid, 1987

Table 4. Prevalence of hookworm infections in primary-school children in different areas of Indonesia

Area	Year	Prevalence (%)	Reference
Jakarta	1986		Margono et al., 1986
Slum		2.4–10.6	
Rural		6.9–29.4	
Suburban	1989	0.6	Margono et al., 1989
South Sumatra	1998	38.8 and 39.7	Margono et al., 1998
West Sumatra	1999		Agus et al., 1999[a]
Municipal		2.5 and 5.2	
District		5.3 and 8.6	
South Sulawesi	1999		Hadju, 1999[b]
Slum		1.4	
Rural		0.5	

[a] Agus Z et al. (1999) *The effectiveness of mass treatment of soil-transmitted helminthiases among school children in West Sumatra.* Report of meeting with Bappenas (Badan Perancang Pembangunan Nasional—the National Planning Board).

[b] Hadju V (1999) *Dampak intervensi pemulihan kecacingan di wilayah PMT-AS Sulawesi Selatan.* Report for Bappenas (Badan Perancang Pembangunan Nasional—the National Planning Board).

children. However, their results—in terms of lowering prevalences of soil-transmitted helminth infections—were not consistent. In Jakarta, for example, the local government has a control programme that operates through school health units. Examination of 397 stool samples from 6 schools in central Jakarta found 17.6% positive for *Ascaris* and 12.3% for *Trichuris* eggs; the figure for total soil-transmitted helminth infection, however, was 25.9% (Margono, 1999, unpublished data). In another control programme implemented in Jakarta by Yayasan Kusuma Buana, a non-government organization, which covered 34 primary schools in 1987 and 507 schools by 1999, prevalence of soil-transmitted helminth infections fell from 78.6% to 11.7%. This programme achieved better cooperation with the schools and the community.

Lymphatic filariasis

Lymphatic filariasis is a major disease in both tropical and subtropical areas and currently affects some 120 million people in 114 countries. In Indonesia, the disease occurs throughout the country with a prevalence between 0.5% and 19.6%, although in some remote and less accessible areas prevalences are higher (Directorate General Communicable Disease Control and Environmental Health, 1998). Compared with the examination of blood obtained by finger-prick (for which blood must be collected at night), survey through questionnaires is considered to be a faster and more useful method of obtaining data on disease burden and of finding new endemic areas.

The causative agents of lymphatic filariasis are the mosquito-borne nematodes *Brugia malayi*, *B. timori*, and *Wuchereria bancrofti*. Adult worms can live for years in the lymph nodes and vessels of lymphatic tissues, and often cause acute and chronic illness. The disease is found in urban and rural areas, and affects people of all ages and both sexes, particularly those of low socioeconomic status.

Lymphatic filariasis is sometimes asymptomatic, but clinical filariasis is often found in endemic areas. On the basis of clinical and laboratory examinations, the disease is categorized as asymptomatic microfilaraemia or as acute, chronic, or occult filariasis. Symptoms include painful attacks of lymphadenitis, and obstructive lesions such as hydrocele and elephantiasis. Filarial damage to the lymphatic vessels is exacerbated by secondary infection due to recurrent episodes of adenolymphangitis. The severity of lymphoedema and elephantiasis is aggravated by the presence of bacterial and fungal superinfections.

The vectors of filariasis in Indonesia include a wide variety of mosquitoes from the genera *Culex*, *Mansonia*, *Aedes*, and *Anopheles*. *Culex quinquefasciatus* is the main vector for urban bancroftian filariasis, whereas in rural areas the disease is transmitted by several species of *Anopheles*, *Culex*, and also *Mansonia*. There are two different types of *Brugia malayi*, known as nocturnal periodic and nocturnal sub-periodic. The nocturnal periodic biotype, endemic in rice field areas and transmitted by *Anopheles barbirostris*, is found mainly in Sulawesi. The sub-periodic biotype of *B. malayi* is zoonotic, with domestic cats and monkeys (*Presbytis* sp.) living close to human habitation serving as reservoir hosts. This type is transmitted by many species of *Mansonia* mosquitoes in six provinces—Bengkulu, Jambi, Riau, South Kalimantan, East Kalimantan, and South-East Sulawesi (Widarso, 1997).

Several new diagnostic tests, including an antigen detection test kit, ultrasonography, and antibody detection tests, have been developed in an effort to replace the night blood examination and to detect occult filariasis. However, their use is still confined to research and epidemiological surveys because of certain limitations of the tests themselves. The antigen detection test, for example, is only for bancroftian filariasis, and may not distinguish between past and current infection. Tests involving the PCR (polymerase chain reaction) technique need a well-equipped laboratory with a qualified technician. Application and interpretation of ultrasonography requires particular expertise.

Indonesian national policy for controlling filariasis continues to rely on mass treatment with low-dose diethylcarbamazine (DEC) regularly every week for 40 doses. The dose is 100 mg/week for individuals aged over 10 years and 50 mg/week for children under 10 years. This programme significantly reduced the prevalence of microfilaraemia from 21.6% in 1970 to 3.1% in 1996 (Widarso & Sri Oemijati, 1997). In 1997, the control programme studied the effect of 0.2% DEC-fortified salt distributed for 9 months in four provinces (Jambi, South Kalimantan, Central Sulawesi, and East Nusa Tenggara). This approach was equally effective in reducing the prevalence of microfilaraemia, and there were hardly any side-effects. The major drawback to its national application is that DEC-fortified salt is much more expensive than regular salt and has a slightly bitter taste.

Trematodes

Schistosoma japonicum

Endemic trematode infections are rare in Indonesia. The most important species to cause a health problem is *Schistosoma japonicum*, found in the area of Lake Lindu and Napu Valley, Central Sulawesi. The disease was discovered for the first time by Müller & Tesch (1937) in a 35-year-old man, who died in the hospital at Palu. Autopsy of the patient revealed eggs of *S. japonicum* in tissue sections. The patient was from Tomado, a village in the area of Lake Lindu. This finding prompted a survey by Brug & Tesch (1937), in which eggs were found in 8% of stool samples. In other surveys, Bonne & Sandground (1940) reported a prevalence of 53% among 176 stool samples, and Buck & Uhrman (1956) reported 26%.

In 1972 stool samples in the Lindu area were again examined, 53% of the samples were egg-positive (Hadidjaja et al., 1972). The prevalence decreased to

17% by 1981. In the area of Napu, prevalence was 43% in 1974 and had declined to 37% by 1982. The Department of Health started a control programme in this area in 1982, covering about 12 000 people at risk in 23 villages in Central Sulawesi; the aim of the programme was to reduce the prevalence of *S. japonicum* infection to less than 1%, when it would no longer be a public health problem. Two doses of praziquantel, 30 mg/kg body weight, were distributed annually. The result was a reduction in prevalence from 17% to 2% after six annual treatments in Lindu Valley and from 33.8% to 2.2% after four annual treatments in Napu Valley. Although the chemotherapy was effective, transmission still continued and retreatment was necessary to maintain the low prevalence.

From 1984, the schistosomiasis control programme in Indonesia included treatment with praziquantel once every 6 months for all infected individuals detected through community surveys and by surveillance, health education, and population migration screening. Control of the intermediate snail host (*Oncomelania hupensis*) was achieved by identifying new snail habitats, applying molluscicide at known transmission sites, reducing the snail population and infection rate, and modifying the environment to make it less suitable for the snails (cutting and burning bushes, drainage and filling of water bodies, and changing from wet to dry agriculture). Efforts were made to provide a safe water supply and sanitation, and systematic treatment of *S. japonicum* infection in cattle and other domestic animals formed part of the control programme (WHO, 1993), conducted at that time by the health sector alone.

Agro-technical considerations should also form part of a control programme, and in 1988 a multisectoral programme was implemented in the schistosomiasis area. This was renamed "integrated development in a schistosomiasis area" in an effort to reduce the domination of control efforts by the health sector. The average infection rate in snails was 2.6% in Lindu and 0.2% in Napu in 1990. A new programme—the "Central Sulawesi integrated area development and conservation project"—started in 1998; its main objectives were development of the area to increase the prosperity of the people and conservation of the Lore Lindu National Park. The health sector played a small role in this project.

The most recent assessment of schistosomiasis was carried out in October 1999. Of 2378 stool samples collected from seven villages in the Lake Lindu area, 0.46% were egg-positive; 1.55% of 4312 samples from 13 villages in the Napu area were positive. Snails were collected in both areas; 6.99% of 415 snails from Lake Lindu and 1.17% of 513 from the Napu area were positive. Among 68 rats collected from Lindu and 77 collected from Napu, 5.88% and 0.00%, respectively, were positive. Stool examinations on other animals (202 and 424) in the two areas revealed 0.00% and 5.83%, respectively, to be egg-positive. The programme is still under way and final evaluation remains to be done (Sri Oemijati, 2000).

Fasciolopsis buski

The intestinal trematode *Fasciolopsis buski* was found to be endemic in South Kalimantan when an 11-year-old boy, from District Babirik, vomited worms, subsequently identified as *F. buski*. Many eggs were found in the boy's stool samples (Hadidjaja et al., 1982). In 1985 a survey conducted in Sei Papuyu village in Babirik revealed that 27.0% of the stool samples collected from 548 people in the community were egg-positive, with the highest prevalence (56.8%) among school-age children (5–14 years). Among individuals aged more than 15 years, prevalence was 9.9%, while in children under 5 years it was 16.4%. The ratio of infected males to females was 1.4 : 1, and most of the infected children were underweight. Complaints were mainly of diarrhoea, colic or heartburn, and loss of appetite; there was vomiting in some cases. All infected individuals were emaciated and anaemic and had distended, tender abdomens. Treatment was with praziquantel, 30 m/kg body weight, divided into two doses. Mild side-effects were recorded in three children: one had headache and two suffered from abdominal discomfort. Four 24-hour faecal specimens were collected and an average of 22 worms were found in each specimen (Handoyo et al., 1986a).

In 1986, a survey in Pajukungan Hulu village found that 7% of stool samples collected from the community were egg-positive. In Parupukan, Teluk Limbang, and Murung Kupang villages, stool samples were collected only from children under 15 years of age; on examination, 56 of 82 (68.3%), 21 of 59 (35%), and 0 of 26 (0%) samples, respectively, proved to be egg-positive. To date, no specialized studies have been conducted on the complete life cycle of *F. buski* in this area. However, it was suspected that metacercariae could be found on water plants, such as water hyacinth (*Eichornia crassipes*), water morning glory (*Ipomoea aquatica*), and lotus (*Nymphoea lotus*), which occur abundantly in the river and swamps near the villages. No studies have yet been done of the first intermediate snail hosts, which possibly include *Gyraulus*, *Lymnaea*, *Indoplanorbis*, *Pila*, and *Viviparus* species, commonly found in the river and in the swampy areas (Handoyo et al., 1986b).

Cestodes

The most common cestodes in Indonesia are the tapeworms *Taenia saginata* and *T. solium*, both of which are endemic in areas where dietary habits and poor standards of sanitation ensure the transmission of infection. Limited surveys and control programmes have been conducted; cases of infection have been reported from time to time.

Since the eggs of these worms cannot readily be distinguished from each other morphologically, surveys based on examination of stool samples provide combined prevalence data for *T. saginata* and *T. solium* in areas such as Bali where both are found. It is also possible that the combined prevalence data include *T. saginata asiatica* as well, which is a subspecies of *T. saginata*. Until now, only *T. solium* has been found in Irian Jaya, whereas *T. saginata asiatica* occurs in North Sumatra. In Bali, both *T. solium* and *Taenia saginata asiatica* have been identified.

Taenia saginata

Between 1962 and 1970, six of the 16 cases of taeniasis diagnosed were found to be due to *T. saginata*; the organism was identified by the Department of Parasitology at the University of Indonesia's Faculty of Medicine in Jakarta. In 1988, seven cases *T. saginata* infection were detected at the same laboratory. Since then, only one or two cases have been documented each year (Margono, unpublished data).

Of 285 individuals interviewed in North Sumatra, 9.5% were found to be positive for taeniasis; the infection was contracted by eating raw or undercooked pork. Later, Fan et al. (1990) confirmed that taeniasis on Samosir Island, North Sumatra, was due to *T. saginata asiatica*.

A survey in Bali found 2.2% of 548 stool samples contained *Taenia* eggs and/or proglottids. The samples were collected from Trunyan, Sukawati, and Padangsambian villages. The infected individuals were aged between 18 months and 60 years (Simanjuntak et al., 1997). In a more recent survey in Banjar Kelod, Bali, examination of 515 stool sample revealed 37 cases of taeniasis, with the infection being 1.7 times more common in males than in females. Most of the infected people were in the age group 20–29 years and had had the infection from between 1 month and 20 years; only 21.2% of them were trying to obtain treatment. The lifestyle and eating habits in Bali were the main factors in the high prevalence of the disease: inhabitants of the island were in the habit of eating "lawar", a dish containing raw pork or other meat mixed with vegetables. Moreover, standards of sanitation were low; people defecated indiscriminately and pigs were found roaming around in the villages (Sutisna, 1990). In 1992, the Department of Health reported 213 cases of taeniasis, which were treated with praziquantel (Sutisna, 1993).

Taenia solium and cysticercosis

The first report of *Taenia* in Indonesia came from Irian: *Taenia* eggs were found in 9% of the stool samples from hospitalized patients in Enarotali. All patients were probably infected with *T. solium*, because the local people kept only pigs; the few cows in the area were owned by newcomers (Tumada & Margono, 1973). Later, a survey in Obano, a village near Enarotali, detected *Taenia* eggs in 2% of 350 stool samples (Margono et al., 1979). At that time an outbreak of seizures and burns occurred in this area, around the Paniai Lakes. The burns were suspected to have been caused by epileptic seizures during the night, when people were sleeping around their fireplaces at the centre of their houses (Subianto, Tumada & Margono, 1978). During 1991, 95 cases of epileptic seizure were reported from Jayawijaya District, east of Paniai District. The disease seemed to be spreading to the east of Irian Jaya, as people moved with their pigs for the purpose of trading. Indiscriminate defecation results in contamination of the soil by *T. solium* eggs, and pigs that roam freely become infected by ingesting eggs from the soil. People then become infected by consuming raw food, especially sweet potato (*Ipomoea batatas*), contaminated with *T. solium* eggs (Simanjuntak et al., 1997), and will suffer from cysticercosis or neurocysticercosis (caused by the development of cysticerci in the brain).

During the years 1991–1995, further reports of epileptic seizures came from Assologaima Subdistrict, Jayawijaya District. The number of cases increased each year, and 13 cases were fatal. Data were collected from the local health centre of Assologaima; the facilities at this centre were relatively good and the outpatient clinic was visited by many people. Most of those who suffered epileptic seizures were aged 15–44 years, but two cases in 1992 and four in 1995 were in the 0–4-year age group. Epileptic seizures were a major cause of mortality. Diagnosis of cysticercosis was based on the histopathological characteristics of cysts resected from people with subcutaneous nodules and from a pig, and was confirmed by mitochondrial DNA analysis. Eighteen serum samples from people with anamnesis of epileptic seizures and 31 from those with subcutaneous nodules were analysed with the immunoblot; 12 of the 18 (67%) and 20 of the 31 (65%) were serologically positive for cysticercosis (Wandra et al., 1999).

Sutisna (1993) reported on six cases of cysticercosis in Bali between 1991 and 1993. Five of the patients were male and one was female; all were aged between 12 and 39 years. One of the male patients had multiple cysts with epileptic seizures, another had multiple cysts but no seizures. The other four patients each had a single cyst without seizures.

Conclusions

The wide distribution of soil-transmitted helminthiases creates difficulties for the implementation of national control activities. Several methods of control should be developed to deal with the different situation in different areas of the country.

Although there is now better control of *S. japonicum* infection in Central Sulawesi, improved diagnostic procedures for detecting infected cases and new foci infection are needed if the situation in this area is to be accurately evaluated. Further investigation of fasciolopsiases in Kalimantan is essential.

Taenia solium and *T. saginata asiatica*, as well as the more common *T. saginata*, are infecting communities in several areas of Indonesia. Further studies of these cestodes are needed to provide a clear picture of the effect of these helminths on the health of people in these regions.

Acknowledgements

The author wishes to express his appreciation to Professor Dr Sri Oemijati and Dr Agnes Kurniawan for the advice they provided during the preparation of this article.

References

Bonne C, Sandground JH (1940) *Bilharzia japonicum* aan het Lindoe meer. [*Bilharzia japonicum* in Lake Lindu.] *Geneeskundig Tijdschrift Nederlandsch-Indie*, 80:477–481.

Brug SL, Tesch JW (1937) Parasitaire wormen aan het Lindoe meer (o.a. Palo, Celebes). [Parasitic worms in Lake Lindu (Palo, Celebes).] *Geneeskundig Tijdschrift Nederlandsch-Indie*, 77:2151–2157.

Carney WP et al. (1974) Intestinal parasites and malaria in Sukomenanti, Pasaman Regency, West Sumatra. *Bulletin Penelitian Kesehatan*, 2:61–66.

Carney WP et al. (1975) Intestinal and blood parasites of man in Timor. *Bulletin Penelitian Kesehatan*, 3:11–15.

Cross JH et al. (1976) Parasitic infections in humans in West Kalimantan (Borneo), Indonesia. *Tropical and Geographical Medicine*, 28:121–130.

Directorate General Communicable Disease Control and Environmental Health, Sub-Directorate Filariasis (1998) *Workshop on Improvement of Filariasis Control Programme in Indonesia*. Bandung, Indonesia.

Fan PC et al. (1990) Indonesia: Taenia and taeniasis. *Yonsei Reports on Tropical Medicine*, 21:33–37.

Hadidjaja P et al. (1972) *Schistosoma japonicum* and other intestinal parasites of inhabitants of Lake Lindu, Sulawesi (Celebes): a preliminary report. *Southeast Asian Journal of Tropical Medicine and Public Health*, 3:594–599.

Hadidjaja P et al. (1982) First autochthonous case of *Fasciolopsis buski* infection in Indonesia. *American Journal of Tropical Medicine and Hygiene*, 31:1065.

Handoyo I et al. (1986a) A survey of fasciolopsiasis in Sei Papuyu village of Babirik Subdistrict, Hulu Sungai Utara Regency, South Kalimantan Province. *Tropical Biomedicine*, 3:113–118.

Handoyo I (1986b) Further survey of fasciolopsiasis in Babirik Subdistrict, Hulu Sungai Utara Regency, South Kalimantan Province. *Tropical Biomedicine*, 3:63–73.

Ismid IS (1996) Infeksi cacing yang ditularkan melalui tanah pada anak balita yang kurang kalori protein di Kelurahan Kramat, Jakarta Pusat. *Majalah Parasitologi Indonesia*, 9:1–5.

Ismid IS, Margono SS, Abidin SAN (1996) Pengaruh pemberian antelmintik terhadap perkembangan telur *Trichuris trichiura*. *Majalah Parasitologi Indonesia*, 9:61–66.

Ismid IS et al. (1999) Infeksi cacing yang ditularkan melalui tanah pada murid sekolah dasar peserta program pemberian makanan tambahan anak sekolah di Mataram, Lombok, Nusa Tenggara Barat. *Majalah Parasitologi Indonesia*, 12:1–9.

Margono SS (1986) *The implementation of intestinal helminths control in an integrated programme in Jakarta* [in Indonesian]. National Consultative Meeting on Diarrhoeal Diseases and Helminthiasis. Bandung, Indonesia.

Margono SS, Ismid IS (1987) Infeksi cacing usus dan pemakaian sarana sanitasi anak balita di Joglo, Jakarta Barat. *Majalah Kesehatan Masyarakat Indonesia*, 678–681.

Margono SS, Ismid IS, Rukmono B (1989) Effect of control of soil-transmitted helminth infections in a suburban area in Jakarta, Indonesia. In: *Collected papers on the control of soil-transmitted helminthiases*, 4:95–104. Tokyo, Asian Parasite Control Organization.

Margono SS et al. (1979) Intestinal parasites in Obano, Irian Jaya, Indonesia. *Majalah Kedokteran Indonesia*, 29:56–58.

Margono SS et al. (1998) *Prevalence and intensity of soil-transmitted helminth infections in two sub-villages of South Sumatra, Indonesia. Association with economic level, behaviour and environmental sanitation*. Presented at 20th Asian Parasite Control Organization, Parasitologists' Meeting, Chiba, Japan.

Margono SS et al. (2000) Intestinal helminthic infections in primary school children in Matraman, Jakarta. *Majalah Kesehatan Masyarakat Indonesia* (in press).

Müller H, Tesch JW (1937) Autochthone infectie met *Schistosoma japonicum* van Celebes. [Autochthonous infection with *Schistosoma* japonicum in the Celebes.] *Geneeskundig Tijdschrift Nederlandsch-Indie*, 77:2143–2150.

Simanjuntak GM et al. (1997) Taeniasis/cysticercosis in Indonesia as an emerging disease. *Parasitology Today*, 13:321–323.

Sri Oemijati. (2000) *Report on schistosomiasis for the Ministry of Health.*

Subianto DB, Tumada LR, Margono SS (1978) Burns and epileptic fits associated with cysticercosis in mountain people of Irian Jaya. Tropical and Geographical Medicine, 30:275–278.

Sutisna P (1990) Masalah taeniasis di Banjar Kelod, Renon, Denpasar. *Medika*, 16:543–549.

Sutisna P (1993) Sistiserkosis di Bali : Laporan 6 kasus. National Seminar Parasitology VII, Bali.

Tumada LR, Margono SS (1973) Intestinal helminthic infection in the Paniai Highlands, with special reference to *Taenia* and *Hymenolepis nana*. *Journal of the Indonesian Medical Association*, 103–107.

Von Buck AA, Uhrmann, G (1956) Untersuchungen über den Befall mit *Echinostoma lindoense* und *Schistosoma japonicum* bei der Bevolkerung des Lindusees (Mittel Celebes). *Zeitschrift für Tropenmedizin und Parasitologie*, 7:110–116.

Wandra T et al. (1999) Resurgence of cases of epileptic seizures and burns associated with cysticercosis in Assologaima, Jayawijaya, Irian Jaya, Indonesia, 1991–95. *Transactions of the Royal Society of Tropical Medicine and Hygiene*, 8:1–5.

Widarso (1997) *Filariasis control in Indonesia*. WHO Intercountry Workshop, Pondicherry, India.

Widarso, Sri Oemijati (1997) *Filariasis control in Indonesia*. 2nd International Conference on the Control of Lymphatic Filariasis, Australia.

WHO (1993) *The control of schistosomiasis. Second report of the WHO Expert Committee.* Geneva, World Health Organization (WHO Technical Report Series, No. 830).

Schistosomiasis

D. Engels,[1] L. Chitsulo
Parasitic Diseases and Vector Control
Communicable Disease Control, Prevention and Eradication
World Health Organization
Geneva, Switzerland

The current status of schistosomiasis and its control

Schistosomiasis remains one of the most prevalent parasitic infections in the world. It is endemic in 76 countries and territories, and continues to be a global public health concern in the developing world. Because it is a chronic insidious disease, it is poorly recognized at early stages, and becomes a threat to development by disabling men and women during their most productive years. It is particularly linked to agricultural and water development schemes and is typically a disease of the poor who live in conditions that favour transmission and have no access to proper care or effective prevention measures.

Although the distribution of schistosomiasis has changed over the past 50 years and there have been successful control programmes, the number of people estimated to be infected or at risk of infection remains unchanged. Where control has been successful, the number of people infected and at risk of infection is very small. This is the situation in most formerly endemic countries in the Americas and in Asia and. In sub-Saharan Africa, on the other hand, where there have been few control programmes and the population has increased by approximately 70% over the past 25 years, a great number of people are infected or at risk of infection.

There are few accurate data on country-specific prevalence of schistosomiasis, and global estimates of the number of people infected and at risk of infection must still be based on extrapolations from the limited prevalence survey data at country level. The most accurate data may be those that derive from national control programmes or national surveys. National surveys conducted in a few countries in Africa have provided estimates of national prevalence. When African prevalence data were extrapolated from the *Atlas of schistosomiasis* (Doumenge et al., 1987) and applied to 1995 population estimates, it was calculated that about 652 million people are at risk of infection from the five human schistosome species

---

[1] Dr D. Engels, Parasitic Diseases and Vector Control, Communicable Disease Control, Prevention and Eradication, World Health Organization, 1211 Geneva 27, Switzerland
Tel: +41 22 791 3824, Fax: +41 22 791 4869, E-mail: engelsd@who.int

and that 193 million people are infected. These same calculations indicate that 85% of the estimated number of infected people are on the African continent (Chitsulo et al., 2000).

The public health impact of schistosomiasis: a growing discrepancy between sub-Saharan Africa and the rest of the world

It cannot be denied that some progress has been made in schistosomiasis control. A number of countries have appreciated the public health importance of schistosomiasis and have initiated control activities. In China and Japan, control efforts were prompted by the high morbidity and mortality due to *Schistosoma japonicum*, which was leading to the disintegration of communities and consequent reduction in agricultural production (Mao & Shao, 1982; Chen, 1989; Tanaka & Tsuji, 1998; Chen & Zheng, 1999). In Brazil, schistosomiasis was among the three major public health problems (Machado, 1982). Control activities were initiated in Egypt, where irrigation is the mainstay of agriculture, because morbidity due to schistosomiasis was thought to be an obstacle to production (Mobarak, 1982). In Morocco, too, the intensive development of irrigated agriculture, and the associated threat of a growing schistosomiasis problem, was the incentive to initiate national control (Laamrani et al., 2000).

Some endemic countries, such as Brazil, China, Egypt, and the Philippines, have been able to sustain national control programmes for a prolonged period and have succeeded in reducing morbidity to very low levels. Others, such as the Islamic Republic of Iran, Mauritius, Morocco, Puerto Rico, Tunisia, Venezuela, and the smaller Caribbean islands, are nearing elimination or have already achieved this goal.

Despite these encouraging results, there is currently little or no schistosomiasis control in sub-Saharan Africa. During the 1980s, a number of African countries implemented donor-funded, vertical control programmes, but these have subsequently proved to be unsustainable. Most national health authorities in sub-Saharan Africa now show little commitment to schistosomiasis control; they may consider that the resources required to control the disease are disproportionate to the size of the public health problem. Despite the numerous reports of substantial late-stage morbidity published in the international literature before 1970, it has been argued that schistosomiasis-related morbidity in sub-Saharan Africa is lower than that in the rest of the world (Gryseels, 1989; Gryseels & Polderman, 1991). The deteriorating socioeconomic situation, a progressive loss of diagnostic capability in peripheral hospital facilities, and the appearance or re-emergence of more obvious health problems may have contributed to the downgrading of schistosomiasis on the public health agenda. The fact that schistosomiasis is a focal problem, which "dilutes" its public health importance at the national level, may have also contributed to the loss of commitment among national health authorities to control of the disease. The net result is that pockets of high morbidity clearly exist in many parts of sub-Saharan Africa and that essential anti-schistosomal drugs are rarely available in these areas.

New prospects for schistosomiasis control

The main principles of schistosomiasis control, such as the concept of morbidity control and its implementation through the primary health care system, have not changed since the 1991 meeting of the WHO Expert Committee (WHO, 1993). However, the past few years have seen change in a number of areas.

Praziquantel—the drug of choice for all forms of schistosomiasis—has become significantly cheaper, and several brands of good quality, generic praziquantel are now available. The average cost of treatment with this drug has fallen to less than a customary user fee, which has clear implications for wider availability of praziquantel. It also implies that both presumptive treatment, based on early clinical symptoms, and universal treatment on the basis of epidemiological criteria have become cost-effective in an increasing number of endemic situations (Guyatt et al., 1994; Carabin, Guyatt & Engels, 2000). As praziquantel is a safe drug, it can be provided at the most peripheral levels of the drug delivery system.

Morbidity control, infection control, and transmission control are increasingly believed to be different objectives that should be recognized as distinct and consecutive steps in control. There is also some recent evidence that the impact of treatment on morbidity declines with age, and that repeated treatment in the early stages of life has a long-lasting effect on morbidity at a later age. If this is indeed the case, focusing the delivery of regular chemotherapy on the younger age groups would produce maximum benefits and prevent chronic sequelae in adulthood (King, Muchiri & Ouma, 1992; Hatz et al., 1998; Frenzel et al., 1999).

The practical tools for epidemiological assessment and for delivery of regular treatment to high-risk groups, with particular emphasis on school-age children, are being constantly refined to meet the requirements for easy and reliable use by individuals other than health workers (e.g. teachers, volunteers) (Chitsulo, Lengeler & Jenkins, 1995; Montresor et al., 1998, 1999).

Schistosomiasis control in high-transmission areas

In areas of high schistosomiasis transmission—now largely in sub-Saharan Africa, the control objective needs to be adapted to the prevailing public health context and to be seen as morbidity control in the stricter sense. The strategy for implementation of control should be simple, easy, and affordable. The World Health Organization has recently reviewed the strategy for morbidity control in high-transmission areas (WHO, 1998), with emphasis on better targeting of control interventions and more cost-effective and sustainable control strategies. Integration of control into existing health structures, schools, workers' cooperatives, etc. and decentralization of decision-making and delivery are key elements for sustainable control.

The provision of adequate clinical care is an essential component of control within the existing health system. In addition, health services should ensure more active morbidity control and should implement appropriate treatment strategies where this is justified by the epidemiological situation. Community-based treatment should be targeted first to school-age children, a high-risk group that can be reached through the primary school system, in collaboration with the educational sector. Even in areas where school enrolment rates are low, well-designed outreach activities can ensure good coverage. School-based delivery systems should be integrated with other public health interventions, such as control of soil-transmitted helminths, feeding programmes, and micronutrient supplementation, whenever possible. Special occupational groups, such as fishermen and irrigation workers, and communities with high prevalence rates should also have access to regular treatment for schistosomiasis.

Long-lasting improvements in hygiene and sanitation, including provision of safe water and appropriate health education, can do much to enhance the effect of regular chemotherapy. Complementary, integrated control activities, such as environmental management measures, should also be planned with other sectors such as agriculture and water resource development programmes.

Schistosomiasis control in low-transmission areas, and prospects for elimination

In a number of countries where the burden of schistosomiasis was once heavy, it has been shown that sustained control efforts have significantly reduced the associated morbidity and mortality. Where the disease is no longer a public health issue, sustainable transmission control should focus primarily on hygiene and sanitation improvement and environmental management. These measures will also contribute to the achievement of other public health goals, and reduce the risk of resurgence of schistosomiasis. As the level of endemic disease declines, new objectives need to be defined with a view to possible elimination. This, in turn, leads to new approaches and algorithms defined according to local situations. In areas of low transmission, cost-effectiveness and decentralized decision-making are crucial to the optimal use of resources and the maintenance of sufficient control to avoid resurgence.

Schistosomiasis is not currently considered by WHO as a disease targeted for eradication or elimination.[1] The Organization has therefore not established a standardized Certification Process, which would involve the establishment of an international commission and of standardized criteria to certify that schistosomiasis were no longer endemic in a country or area. The definition of criteria for elim-

[1] *Eradication* is defined as the permanent reduction to zero of the worldwide incidence of a disease as a result of deliberate efforts; continued intervention measures are no longer required.
Elimination is defined as the reduction to zero of the incidence of a specific disease in a defined geographical area as a result of deliberate efforts; continued intervention measures are repuired.

ination would in any case be a complex issue given that schistosomiasis has a complex transmission cycle, that the asymptomatic carrier state is common, and that for certain parasite species an animal reservoir exists. Moreover, interruption of transmission may be achieved in different ways: by "sterilization" of the para site reservoir; by elimination of the snail intermediate host (e.g. through the use of competitor snails); by improvements in socioeconomic status and hygiene to prevent contamination of water, and subsequent contact with contaminated water; or by a combination of these measures. There is also the risk of reintroduction of the disease into an area from which it has previously been eliminated, particularly where there are water resource development activities or significant population migration.

Nevertheless, it is possible for individual countries to demonstrate that they have eliminated schistosomiasis, by documenting that no new, locally contracted infections have been observed over a specified period of time. The observation period necessary to validate a claim that transmission has been interrupted depends greatly on the risk of re-emergence or reintroduction in the particular context. Similarly, the degree of certainty that no new cases have been detected depends on the performance of the surveillance system, in terms of sensitivity of the diagnostic method used and the reliability of the reporting system. Ideally, close surveillance should be implemented through existing, permanent channels and procedures, complemented by surveys in populations at high risk. Even in low-transmission areas, older primary-school children and special occupation groups appear to be the best populations among which to conduct surveillance surveys. Surveillance should be eased only when the risk of resurgence can be clearly demonstrated to have diminished (WHO, 2001).

Integration of schistosomiasis control into other public health interventions: multi-disease approach

The integration of schistosomiasis control into existing health care delivery structures and public health interventions, such as school health packages, is essential in high-transmission areas in order to ensure health policy commitment and sustainability. Health authorities should recognize schistosomiasis control as an integral part of health packages delivered to the populations for which they have responsibility.

Many diseases are intimately linked to poverty and poor living conditions. This is particularly true for parasitic diseases, and poor people are therefore likely to have more than one parasitic disease. For many of these diseases, regular chemotherapy is the means of relieving the disease burden in the short term, while improved living conditions and economic development are the permanent solutions.

There are sound reasons for linking schistosomiasis control with the control of soil-transmitted helminthiasis. The ecological conditions that favour transmission of the two diseases are similar, resulting in a wide geographical overlap of their distribution. In both cases, the aim of control in high-transmission areas is firstly to reduce morbidity and thereafter to keep it at a low level, which can be achieved by the delivery of regular chemotherapy to high-risk groups, particularly school-age children. Since chemotherapy alone has only a limited impact on the transmission of both schistosomiasis and soil-transmitted helminthiasis, more permanent control of both conditions requires substantial improvements in water supplies and sanitation, reinforced by appropriate health education.

An integrated, multi-disease approach of this type, which may also be extended to other parasitic diseases, will maximize the use of infrastructure, personnel, and resources, as well as the health benefits for the target populations. In areas where some parasitic diseases have been successfully controlled, the experience gained and infrastructure used in the process can become the foundation for other public health initiatives. In defining and delivering the most appropriate and cost-effective interventions, it is important to take into account disease-specific elements such as control objectives (morbidity control, transmission control, or elimination), epidemiological distribution, and optimal delivery and timing of interventions. Clearly, the implementation of an extended multi-disease approach in a given situation may be complex, and more practical experience is required in this field before formal strategies can be formulated.

Finally, it should be stressed that, since the principal underlying causes of the commonest parasitic diseases are poor hygienic conditions, integration of control efforts with services dealing with hygiene-related diseases should have a synergistic effect in reducing poverty-related diseases in general.

Conclusions

The control of schistosomiasis, a disease that still affects many poor people in the developing world, deserves more and renewed attention and commitment, particularly in sub-Saharan Africa. Simple but sustained control measures can relieve an unnecessary—and underestimated—disease burden in areas of high transmission. This has been demonstrated in a number of countries where sustained control efforts mean that elimination of the disease can now be realistically contemplated.

Schistosomiasis control is not an "all-or-nothing" phenomenon, and a few basic measures can easily be implemented in all circumstances. Essential anti-schistosomal drugs must be accessible at all levels of the health system, and appropriate treatment strategies must be established in accordance with the endemic level. Protecting the younger members of the population from subtle morbidity and late-stage complications is within the reach of most countries, with or without the help of international donor agencies, particularly if regular anthelminthic drug treatment is integrated into a comprehensive health package delivered to school-age children. More substantial and costly control measures may be adopted if a country wants to extend schistosomiasis control beyond reduction of morbidity, provided that the technical efforts can be sustained and financed over sufficiently long periods.

References

Carabin H, Guyatt H, Engels D (2000) A comparative analysis of the cost-effectiveness of treatment based on parasitological and symptomatic screening for *Schistosoma mansoni* in Burundi. *Tropical Medicine and Parasitology*, 5:192–202.

Chen MG (1989) Schistosomiasis control program in the People's Republic of China: a review. *Southeast Asian Journal of Tropical Medicine and Public Health*, 20:511–517.

Chen MG, Zheng F (1999) Schistosomiasis control in China. *Parasitology International*, 48:11–19.

Chitsulo L, Lengeler C, Jenkins J (1995) *The schistosomiasis manual. A guide for the rapid identification of communities with a high prevalence of urinary schistosomiasis.* Geneva, World Health Organization (document TDR/SER/MSR/95.2).

Chitsulo L et al. (2000) The global status of schistosomiasis and its control. *Acta Tropica*, 77:41–51.

Doumenge JP et al. (1987) *Atlas de la répartition mondiale des schistosomiases. [Atlas of global distribution of schistosomiasis.]* Bordeaux, Presses Universitaires de Bordeaux.

Frenzel K et al. (1999) Evidence for a long-term effect of a single dose of praziquantel on *Schistosoma mansoni*-induced hepatosplenic lesions in northern Uganda. *American Journal of Tropical Medicine and Hygiene*, 60:927–931.

Gryseels B (1989) The relevance of schistosomiasis for public health. *Tropical Medicine and Parasitology*, 40:134–142.

Gryseels B, Polderman A (1991) Morbidity due to schistosomiasis mansoni, and its control in sub-Saharan Africa. *Parasitology Today*, 7:244–248.

Guyatt H et al. (1994) Controlling schistosomiasis: the cost-effectiveness of alternative delivery strategies. *Health Policy and Planning*, 9:385–395.

Hatz CF et al. (1998) Evolution of *Schistosoma haematobium*-related pathology over 24 months after treatment with praziquantel among school children in southeastern Tanzania. *American Journal of Tropical Medicine and Hygiene*, 59:775–781.

King CH, Muchiri EM, Ouma JH (1992) Age-targeted chemotherapy for control of urinary schistosomiasis in endemic populations. *Memorias do Insituto Oswaldo* Cruz, 87:203–210.

Laamrani H et al. (2000) *Schistosoma haematobium* in Morocco: moving from control to elimination. *Parasitology Today*, 16:257–260.

Machado PA (1982) The Brazilian Program for schistosomiasis control, 1975–1979. *American Journal of Tropical Medicine and Hygiene*, 31:76–86.

Mao S, Shao B (1982) Schistosomiasis control in the People's Republic of China. *American Journal of Tropical Medicine and Hygiene*, 31:92–99.

Mobarak AB (1982) The schistosomiasis problem in Egypt. *American Journal of Tropical Medicine and Hygiene*, 31:87–91.

Montresor A et al. (1998) *Guidelines for the evaluation of soil-transmitted helminthiasis and schistosomiasis at community level.* Geneva, World Health Organization (document WHO/CTD/SIP/98.1).

Montresor A et al. (1999) *Monitoring helminth control programmes at community level. Guidelines for monitoring the impact of control programmes aimed at reducing morbidity caused by soil-transmitted nematodes and schistosomes, with particular reference to school-age children.* Geneva, World Health Organization (document WHO/CDS/CPC/SIP/99.3).

Tanaka H, Tsuji M (1997) From discovery to eradication of schistosomiasis in Japan: 1847–1996. *International Journal of Parasitology*, 27:1465–1480.

WHO (1993) *The control of schistosomiasis. Second report of the WHO Expert Committee.* Geneva, World Health Organization (WHO Technical Report Series, No. 830).

WHO (1998) *Report of the Informal Consultation on Schistosomiasis Control, Geneva 2–4 December 1998.* Geneva, World Health Organization (document WHO/CDS/CPC/SIP/99.2).

WHO (2001) *Report of the Informal Consultation on schistosomiasis in low transmission areas: control strategies and criteria for elimination, London, 10–13 April 2000.* Geneva, World Health Organization (document WHO/CDS/CPE/SIP/2001.1).

Hookworm infection and iron status

D.W.T. Crompton,[1] H. Torlesse, M.E. Hodges
WHO Collaborating Centre for Soil-transmitted Helminthiasis
Institute of Biomedical and Life Sciences
University of Glasgow, Glasgow, G12 8QQ, Scotland

Introduction

Lines of circumstantial evidence, including early writings, the medical treatises of Hippocrates and Lucretius, and examination of coprolites, indicate that humans have been infected with hookworms for centuries (Ball, 1996). Adult *Ancylostoma duodenale* were found in the small intestine of a mummy from a burial site in Tiahuanaco, Peru, dating from about A.D. 890–950 (Vreeland & Cockburn, 1980). Informed speculation suggests that hookworms reached the Americas with the early migrants who crossed the Bering land bridge. Today, *Necator americanus* and *A. duodenale* occur together throughout the tropics and subtropics; *A. duodenale* also survives in cooler and drier climates such as those of northern China, north-west India, North Africa, southern Europe, and the eastern Mediterranean countries (Pawlowski, Schad & Stott, 1991). An estimated 1.3 billion people are currently infected with hookworms (Crompton, 1999).

The intestinal stages of both species of hookworm feed on blood and cause further haemorrhage when they stop feeding because they release anticoagulant compounds (Hotez & Cerami, 1983). Secure evidence for the fact that hookworm infection can be a determinant of iron-deficiency anaemia was not obtained until the late 19th century when the construction of the St Gotthard tunnel between Italy and Switzerland was disrupted by illness among the miners (Ball, 1996). Many miners died from the consequences of severe anaemia. Although the route of hookworm transmission was not known, the problem was overcome by improving hygiene and disposing of the workers' excreta. Hookworm disease would largely disappear if the populations of areas where hookworms persist had access to appropriate and affordable sanitation.

Details of the life history (Hoagland & Schad, 1978) and population biology (Anderson & May, 1991) of hookworms are now well known. The infective larvae of *N. americanus* have an obligatory need to penetrate skin, but larvae of *A. duodenale* are capable of infecting humans not only by skin penetration but also when

[1] Professor D.W.T. Crompton, WHO Collaborating Centre for Soil-transmitted Helminthiasis, Institute of Biomedical and Life Sciences, Graham Kerr Building, University of Glasgow, Glasgow G12 8QQ, Scotland
Tel: +44 1838 400203, Fax: +44 1301 702430, E-mail: dwtc@tyndrum.demon.co.uk

swallowed. Recent evidence indicates that *A. duodenale* can cross the placenta and establish in the fetus (Sen-Hai & Wei-Xia, 1990).

The purpose of this brief review is to consider the importance of hookworm infection as a factor in impaired iron status and iron-deficiency anaemia. Morbidity caused by the other common species of soil-transmitted helminth is often difficult to diagnose and quantify. In the case of hookworms, however, authorities seem to agree that hookworm disease can be described as hookworm-related iron-deficiency anaemia and associated ill-health (Pawlowski, Schad & Stott, 1991).

Iron requirements during the human life cycle

Iron is an essential micronutrient that contributes to the production of haemoglobin, the transport of electrons in cells, and the synthesis of a range of enzymes. Since iron is a common element, and ubiquitous in the earth's crust, it is surprising to find that some degree of iron deficiency affects over half the world's population (ACC/SCN, 2000). Iron is also toxic, and traits selected during the course of human evolution have resulted in the conservation of body iron stores and restricted absorption of iron from food being digested in the gut.

Iron requirements vary throughout the human life cycle and during different phases of a particular stage in the life cycle. For example, a woman of childbearing age requires more iron during a pregnancy than she does between pregnancies. The Food and Nutrition Board of the USA regularly evaluates the quantities of nutrients required daily to sustain good health. It has been concluded that the daily intake of iron should be sufficient to maintain an iron store of 300 mg in an adult (NAS, 1989). Obviously, this figure will differ for infants, adolescents, and other age groups (Crompton & Whitehead, 1993).

Maintenance of iron stores is a balance between physiological losses and dietary intake. The recommended daily intake for adult men is 10 mg and for adolescent and adult women 15 mg. These recommendations probably have application beyond the USA, particularly since many Americans are of African descent: it is now recognized that the iron metabolism of Africans is different from that of Caucasians (Perry et al., 1992). Crompton & Whitehead (1993) have provided a detailed discussion of the iron status of the individual and the factors that affect iron balance. In industrialized countries, including the USA, most people obtain iron from animal foods (haem iron), which is far more readily absorbed than iron from vegetable sources (non-haem iron). The majority of people with poor iron status are vegetarians, though not necessarily from choice—any animals in their possession are too important economically to be used for food. Non-haem iron is difficult to absorb (Carpenter & Mahoney, 1992) and is usually a minor component of vegetable foods. For example, 100 g of raw liver contains 11 mg of haem iron, while 100 g of rice contains only 0.5 mg of non-haem iron. Various components of vegetable foods, such as phytates and tannates, reduce the bioavailability of iron for absorption in the gut lumen. It should be noted that hookworm infections are endemic where vegetarian diets are the norm, and recommendations are needed for daily intakes of iron by people who must rely on non-haem iron to sustain their iron stores.

Global iron situation

According to the United Nations Administrative Committee on Coordination, Sub-Committee on Nutrition (ACC/SCN, 2000), about 3.5 billion people worldwide are affected by iron deficiency and many of these present with anaemia. For these purposes, anaemia is defined as a blood haemoglobin concentration below a specified cut-off value for a particular age range and for the sex of the individual (see Table 1). The ACC/SCN figure suggests that the iron situation has deteriorated markedly during the past decade: in 1994, Viteri had estimated that about half of all iron-deficient people were anaemic. The fact remains that the anaemia situation is much more severe in developing than in developed countries; 56% of pregnant women, 53% of school-age children, 44% of non-pregnant women, and 42% of preschool children are judged to have blood haemoglobin values below the cut-off points shown in Table 1. The highest prevalence of iron-deficiency anaemia—75%—occurs in south central Asia.

Determinants of impaired iron status and iron-deficiency anaemia

In industrialized countries, poor iron status and anaemia (Figure 1) are probably accounted for mainly by reduced bioavailability of iron for absorption. In developing countries, the etiology is likely to be much more complex and can arise from a series of social and cultural factors (Sanjur, 1982). For example, the supply of iron-rich foods to individuals in a household will be linked to family spending power, family food distribution habits, religious practices, seasonal influences on food production, methods of food preparation, and the effects of infections.

Among parasitic infections, malaria, trichuriasis, and *Schistosoma haematobium* infection are known to contribute to poor iron status and iron-deficiency anaemia

Table 1. Cut-off points for human blood haemoglobin concentration[a] as a definition of anaemia

Population group	Haemoglobin level (g/litre)
6–59 months	110
5–11 years	115
12–14 years	120
Non-pregnant women	120
Pregnant women	110
Adult males	130

[a] Source: ACC/SCN, 2000.

Figure 1. Prevalence of anaemia by age group in industrialized and developing countries[a]

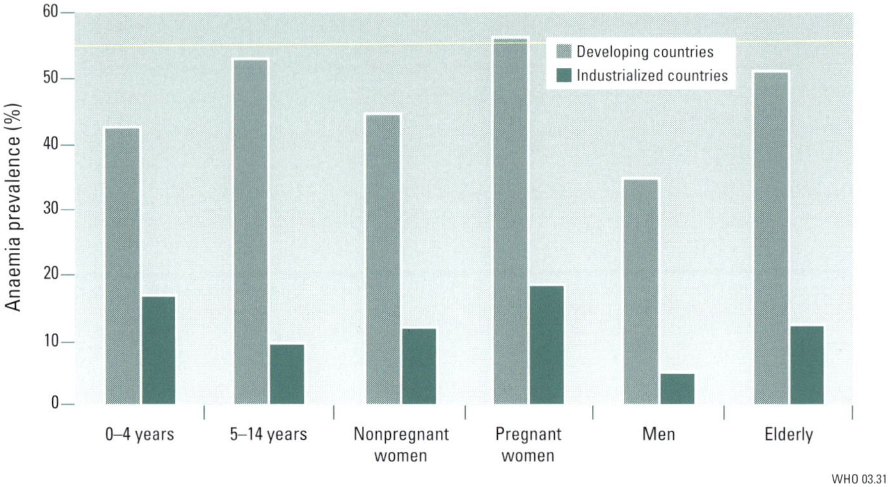

[a] Source: ACC/SCN, 2000

(Stephenson, 1993; Crompton, 2000); these infections, singly or in combination, often occur concurrently with hookworm in the same person. A number of studies have shown that a single intestinal stage of one hookworm is estimated to cause a daily blood loss into the gut of from 0.03–0.15 ml (for reviews and sources of evidence see Crompton & Whitehead, 1993; Crompton, 2000). On the basis of this amount of blood loss, Pawlowski, Schad & Stott (1991) estimated that 25 *N. americanus* cause a daily loss of 0.35 mg of iron into the gut. This iron is in the haem form and some of it may be reabsorbed but, inevitably, some will leave the host in the stools.

Stoltzfus et al. (1996) investigated the relationship between blood haemoglobin concentration and the intensity of hookworm infection in schoolchildren in Zanzibar. The results (Figure 2) demonstrate unequivocally a strong statistical relationship between a fall in blood haemoglobin concentration and increasing intensity of infection due mainly to *N. americanus*. Intensity was measured indirectly by counting the number of eggs passed in the stools of the study subjects. It is generally safe to assume that the greater the egg count, the greater number of female worms present. In a dioecious species such as *N. americanus* or *A. duodenale*, male worms will also be present, probably in similar numbers. In studies during the past 60 years, a direct relationship has been detected between decline in blood haemoglobin concentration and increasing intensity of hookworm infection (Hill & Andrews, 1942; Roche & Layrisse, 1966; Crompton & Stephenson, 1990). The consistency of these observations explains why reducing worm burdens, and thereby morbidity, should be the aim of a helminth control programme when anthelminthic drug treatment provides the main control measure.

Figure 2. The relationship between iron deficiency and the intensity of hookworm infection in schoolchildren from Zanzibar[a]

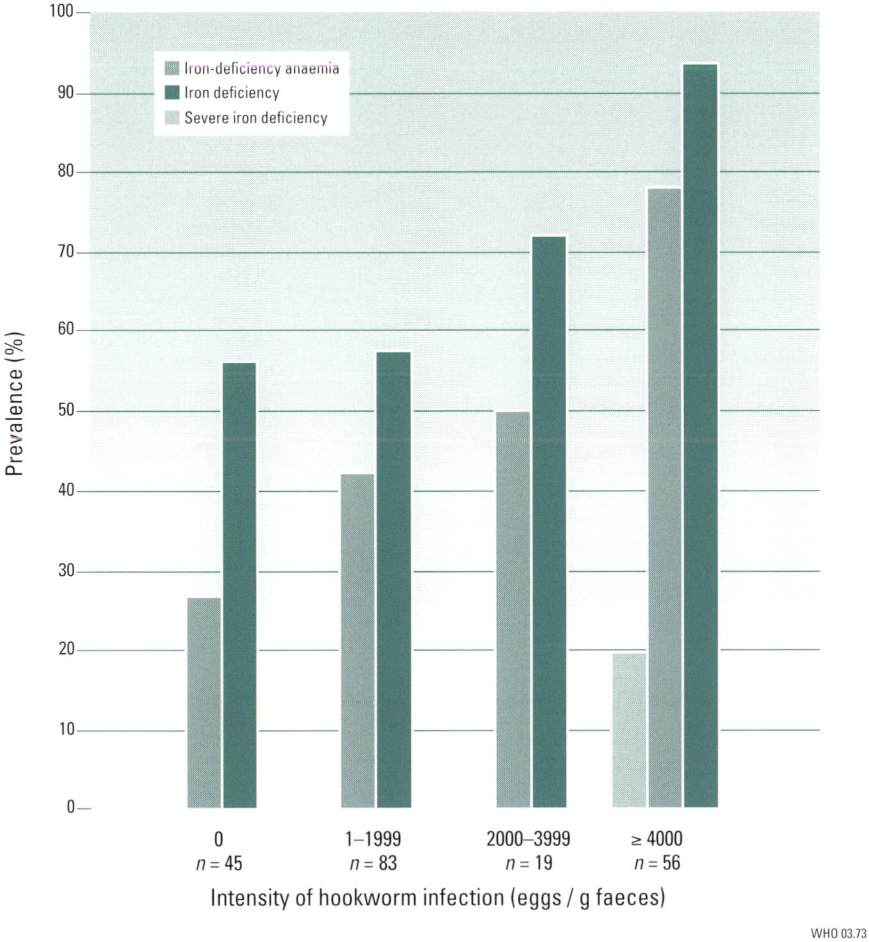

[a] Source: Stoltzfus et al., 1996

Pregnancy and hookworm infection

Concern over the problem of anaemia in adolescent girls and women of child-bearing age, especially in developing countries, led the World Health Organization to review the problem and make recommendations intended to bring about improvements (WHO, 1996). Many women in developing countries spend most of their reproductive years either pregnant or lactating. Each pregnancy requires a transfer from mother to fetus of about 300 mg of iron during the third trimester, the mother requires an additional 500 mg of iron to cope with the increased red blood cell mass needed for a successful pregnancy, and each day of lactation involves a transfer of about 0.75 mg of iron from mother to child. Anaemia during

Table 2. Estimated numbers infected with hookworm and at risk of morbidity in different regions of the world assuming different threshold worm burdens for morbidity

Region[a]	Births	Pregnancies	Infected	Worm burden >100	Worm burden >200
SSA	23320	23553	7537	596	426
LAC	11959	12079	4107	496	327
MEC	13892	14031	3087	171	116
IND	25690	25947	10898	1528	1025
CHN	25065	25316	8607	635	349
OAI	23091	23322	10028	1075	700
Total	123017	124247	44264	4502	2944

Note: Fetal mortality is assumed to be 1%. Estimates are in thousands.
[a] World Bank regions:
SSA sub-Saharan Africa.
IND India.
LAC Latin America and Caribbean.
CHN Peoples' Republic of China.
MEC Middle Eastern Crescent.
OAI Other Asia and Islands.

pregnancy is associated with premature delivery, low birth weight, maternal ill-health, and maternal death (Seshadri, 1997). The problem is so serious that two of the global targets proposed at the World Summit for Children, held in New York in 1990, were to reduce the incidence of low birth weight (defined as 2.5 kg or less) to no more than 10% and to reduce by a third the prevalence of iron-deficiency anaemia in women (World Bank, 1993).

Blood loss caused by hookworm infections will put mother, fetus, and child at risk of iron deficiency leading to anaemia. The extent to which this occurs will depend on host iron status, the infecting species, and the intensity and duration of infection. In developing countries, the iron status of women at conception is frequently poor as a result of inadequate dietary iron intake, concurrent infections, and frequent, closely spaced pregnancies (WHO, 1993). Theoretical analyses suggest that the worm-burden threshold that will precipitate iron depletion and anaemia may be lower during pregnancy when there is increased pressure on iron stores (Crompton & Whitehead, 1993). An inverse relationship between blood haemoglobin concentration and hookworm egg counts has been observed among pregnant women in Liberia (Jackson & Jackson, 1987) and Sierra Leone (Torlesse, 1999).

Estimates of the extent of the hookworm–pregnancy issue are set out in Table 2. A number of countries have included hookworm control as part of their anaemia control programmes in schoolchildren, but very few have promoted routine anthelminthic treatment in pregnant women because of a lack of data on the safety of the drugs during pregnancy. After a thorough review of information about anthelminthic drugs (albendazole, levamisole, mebendazole, and pyrantel) and with due regard to the seriousness of hookworm infection during pregnancy, a WHO Informal Consultation concluded that a single, oral dose of anthelminthic drug could be given to pregnant and lactating women but that, as a general rule, no drug should be given during the first trimester (WHO, 1996).

Two important studies have recently contributed to knowledge of the consequences of using anthelminthic drugs to reduce morbidity during pregnancy. The first, undertaken in Sri Lanka, examined the effect of mebendazole treatment during pregnancy on rates of major congenital defects, stillbirth, perinatal death, and low birth weight (de Silva et al., 1999). No significant difference was observed between the rates of birth defects among infants of mothers who had taken mebendazole during pregnancy and the rates among infants of untreated mothers.

The second study, in western Sierra Leone, examined the role of intestinal nematode infections, including hookworm, in the etiology of iron deficiency and anaemia in pregnant women (Torlesse, 1999; Torlesse & Hodges, 2001). The protocol involved 125 pregnant women who were randomly assigned to treatment groups (albendazole, iron–folate supplements, and their respective placebos) in the second trimester. In the first trimester, the prevalence of hookworm infection (*N. americanus*) was 66.5%. Anaemia was found in 56.0% of the women and iron-deficiency anaemia (based on serum ferritin <20 µg/litre in anaemic women) in 21.2% of the women. Removal of intestinal nematode infections with albendazole combined with daily iron–folate supplements stabilized the prevalence of anaemia and iron-deficiency anaemia and minimized the decline in haemoglobin and serum ferritin concentrations during pregnancy. Torlesse (1999) concluded that pregnant women in this part of Africa would benefit from anthelminthic treatment after the first trimester and regular iron–folate supplements during the first trimester. In the pregnancies and outcomes studied, there was no indication that the anthelminthic drug had affected birth defect rates.

Cognitive development and hookworm infection

Consensus is emerging that iron deficiency impairs the cognitive development of children and, since the learning abilities of children provide the skills for development, will contribute to loss of productivity (ACC/SCN, 2000). Pollitt (1990) has been prominent in studying cognitive development in relation to iron deficiency and has concluded that there is ample evidence to link iron deficiency with impaired educational performance. If poor iron status impairs cognitive development and educational performance, it follows that hookworm infection will contribute to these adverse effects. The problems of investigating cognitive development have been reviewed by Connolly & Kvalsvig (1993).

Physical fitness, worker productivity, and hookworm infection

The events that brought the construction of the St Gotthard tunnel to a standstill and the measures taken by the United Kingdom government to protect its mining industry at the start of the 20th century show that hookworm-induced iron-deficiency anaemia can reduce physical fitness and worker productivity (Ball, 1996). The St Gotthard episode was an extreme case; it is much more difficult to quantify the effects of hookworm infection on the productivity of agricultural workers, building workers, road construction teams, and so on over longer periods of time. A challenging question would be: would the economy and development of a region improve if morbidity due to hookworm infection were to be controlled? Crompton & Stephenson (1990) reviewed literature published to that date which showed that worker productivity improved, to a varying extent, when measures were taken to correct anaemia. Heavy manual work cannot be sustained once the blood haemoglobin concentration falls below 70 g/litre (Fleming, 1982). Iron-deficient road workers in Kenya, rubber tappers in Indonesia, and tea pickers in Sri Lanka showed a diminished work capacity and productivity when compared with healthy subjects (Holland, 1987). Holland concluded that the potential for improved productivity in hookworm-infected adults worldwide must be substantial. The functional consequences of iron deficiency are now being widely recognized. In a recent report from ACC/SCN (2000) we read, "A recent analysis of the economic consequences of iron deficiency has estimated the median value of productivity losses due to iron deficiency to be about US$ 4 per capita, 0.9% of GDP [gross domestic product], for a range of developing countries. The dominant effect is the loss associated with cognitive deficits in children. This estimate does not include the burden of maternal death associated with severe anaemia, nor the lowered effectiveness of funds spent on education."

Conclusions

From this brief survey, which has sought to highlight the relationship between hookworm infection and the worldwide anaemia problem, the case for promoting actions to control morbidity due to hookworm infection seems extremely convincing. Major agencies recognize the health risks to pregnant women caused by iron-deficiency anaemia (WHO, 1996; World Bank, 1999; ACC/SCN, 2000); measures to reduce the impact of hookworm infection must help to reduce the problem of anaemia in pregnancy. There is no clear guidance as to the number of hookworms per person that constitutes a threat to health. The factors that determine whether or when a hookworm burden will cause morbidity include the host's iron stores, the dietary sources of iron, the bioavailability of the ingested iron, concurrent infections, and prevailing cultural attitudes. Hookworm control can readily be integrated into existing health programmes; a mass of experience is available to show how this can be achieved.

References

ACC/SCN (2000) *Fourth report on the world nutrition situation.* Geneva, United Nations Administrative Committee on Coordination, Sub-Committee on Nutrition.

Anderson RM, May RM (1991) *Infectious diseases of humans.* Oxford, Oxford University Press.

Ball PAJ (1996) Hookworm disease. In: Cox FEG, ed. *Illustrated history of tropical diseases*. London, The Wellcome Trust:318–325.

Carpenter CE, Mahoney AW (1992) Contributions of heme and nonheme iron to human nutrition. *Critical reviews in food science and nutrition*, 31:333–367.

Connolly KJ, Kvalsvig JD (1993) Infection, nutrition and cognitive performance in children. *Parasitology*, 107(Suppl.):S187–S200.

Crompton DWT (1999) How much human helminthiasis is there in the world? *Journal of Parasitology*, 85:397–403.

Crompton DWT (2000) The public health importance of hookworm disease. *Parasitology*, 121(Suppl.):S39–S50.

Crompton DWT, Stephenson LS (1990) Hookworm infection: nutritional status and productivity. In: Schad GA, Warren KS, eds. *Hookworm disease*. London, Taylor & Francis:231–264.

Crompton DWT, Whitehead RR (1993) Hookworm infections and human iron metabolism. *Parasitology*, 107(Suppl.):S137–S145.

de Silva NR et al. (1999) Effect of mebendazole therapy in pregnancy on birth outcome. *Lancet*, 353:1145–1149.

Fleming AF (1982) Iron deficiency in the tropics. *Clinics in Haematology*, 11:365–387.

Hill AW, Andrews J (1942) Relation of hookworm burden to physical status in Georgia. *American Journal of Tropical Medicine*, 22:499–502.

Hoagland KE, Schad GA (1978) *Necator americanus* and *Ancylostoma duodenale*: life history parameters and epidemiological implications of two sympatric species. *Experimental Parasitology*, 44:36–49.

Holland C (1987) Hookworm infection. In: Stephenson LS. *Impact of helminth infections on human nutrition*. London, Taylor & Francis:128–160.

Hotez PJ, Cerami A (1983) Secretion of a proteolytic anticoagulant by *Ancylostoma duodenale* hookworms. *Journal of Experimental Medicine*, 157:1594–1603.

Jackson RT, Jackson LC (1987) Biological and behavioral contributors to anemia during pregnancy in Liberia, West Africa. *Human Biology*, 59:585–597.

NAS (1989) *Recommended dietary allowances*, 10th ed. Washington, DC, National Academy Press for the National Academy of Sciences.

Pawlowski ZS, Schad GA, Stott GJ (1991) *Hookworm infection and anaemia*. Geneva, World Health Organization.

Perry GS et al. (1992) Iron nutrition does not account for the haemoglobin differences between blacks and whites. *Journal of Nutrition*, 122:1417–1422.

Pollitt E (1990) *Malnutrition and infection in the classroom*. Paris, United Nations Educational, Scientific and Cultural Organization.

Roche M, Layrisse M (1966) The nature and causes of "hookworm anaemia". *American Journal of Tropical Medicine and Hygiene*, 15:1030–1100.

Sanjur D (1982) *Social and cultural perspectives in nutrition*. Englewood Cliffs, NJ, Prentice-Hall.

Seshadri S (1997) Nutritional anaemia in South Asia. In: Gillespie S, ed. *Malnutrition in South Asia*. Kathmandu, United Nations Children's Fund:75–124.

Sen-Hai Y, Wei-Xia S (1990) Hookworm infection and disease in China. In: Schad GA, Warren KS, eds. *Hookworm disease*. London, Taylor & Francis:44–54.

Stephenson L (1993) The impact of schistosomiasis on human nutrition. *Parasitology*, 107:S107–S123.

Stoltzfus RJ et al. (1996) Hemoquant determination of hookworm of hookworm-related blood loss and its role in iron deficiency in African children. *American Journal of Tropical Medicine and Hygiene*, 55:399–404.

Torlesse H (1999) *Parasitic infection and anaemia during pregnancy in Sierra Leone* [Thesis]. Glasgow, University of Glasgow.

Torlesse H, Hodges M (2001) Anthelminthic therapy and reduced decline in haemoglobin concentration during pregnancy (Sierra Leone). *Transactions of the Royal Society of Tropical Medicine and Hygiene*, 95:195–201.

Viteri FE (1994) The consequences of iron deficiency and anaemia in pregnancy on maternal health, the foetus and the infant. *SCN News*, 11:14–18.

Vreeland JM, Cockburn A (1980) Mummies of Peru. In: Cockburn A, Cockburn E, eds. *Mummies, disease and ancient cultures.* Cambridge, Cambridge University Press:135–174.

WHO (1993) *Prevention and management of severe anaemia in pregnancy. Report of a Technical Working Group.* Geneva, World Health Organization (document WHO/FHE/MSM/93.5).

WHO (1996) *Report of the WHO Informal Consultation on Hookworm Infection and Anaemia in Girls and Women.* Geneva, World Health Organization (document WHO/CTD/SIP/96.1).

World Bank (1993) *World development report 1993.* New York, Oxford University Press.

World Bank (1999) *World development indicators.* Washington, DC, World Bank.

Helminth infections, growth, and anaemia: lessons from Zanzibar

Hababu M. Chwaya
Public Health Laboratory "Ivo de Carneri",
Zanzibar, United Republic of Tanzania

Rebecca J. Stoltzfus[1]
Center for Human Nutrition, Department of International Health,
School of Hygiene and Public Health,
Johns Hopkins University, Baltimore, MD, USA

Over the past six years, we have been engaged in a collaborative research programme with the World Health Organization to determine the most effective strategies for improving the health and development of Zanzibari children through the control of intestinal parasitic infections. This paper explains our initial expectations of the nutritional benefits of helminth control, how research results have both confirmed and expanded those expectations, and our revised expectations of the several ways in which helminth control might affect children's health and development. We focus on the outcomes of growth and anaemia to illustrate the lessons we have learned from Zanzibar.

The studies discussed in this paper were carried out in Pemba Island, which is the more northerly of the two major islands that comprise Zanzibar, within the United Republic of Tanzania. Pemba lies just north of the equator and is extremely densely populated, with an estimated 400 000 people. There is intense transmission of the four soil-transmitted helminths—*Ascaris lumbricoides, Trichuris trichiura*, and the hookworms (Renganathan et al., 1995). Both species of hookworms (*Ancylostoma duodenale* and *Necator americanus*) are found on the island, but *N. americanus* is the more widespread (Albonico et al., 1998). *Schistosoma haematobium* is also endemic, although more focal in transmission than the geohelminths, and has been the focus of a school-based control programme since 1996 (Savioli et al., 1990). *Plasmodium falciparum* malaria is holo-endemic, with year-round transmission and a peak in morbidity in the summer, following the major period of rains. In this context, our goal has been to determine the most efficacious and cost-effective strategies for helminth control, guided by the

[1] Dr Rebecca J. Stoltzfus, Division of Nutritional Sciences, Cornell University, Ithaca, NY 14853, USA (present address)
Tel: +1 607 255 7671, Fax: +1 607 255 1033, E-mail: rjs62@cornell.edu

priority goals of the Ministry of Health to reduce childhood malnutrition and anaemia.

Our initial expectations were guided by the framework shown in Figure 1. The key variable predicting morbidity in this framework is worm burden; that is, we expected that morbidity—in this case, malnutrition—would be directly related to the number of worms inhabiting a child's gut. The important implication of this model is that children with the greatest worm burden will be at greatest nutritional risk and will benefit most from deworming. The idea that worm burden drives helminth-related morbidity has guided global policy on helminth control, and is a major reason why schoolchildren (who harbour high worm burdens) have been prioritized as a target group for anthelminthic treatment programmes (Savioli, Bundy & Tomkins, 1992).

Previous collaborative research by the Ministry of Health and WHO had demonstrated that Zanzibari children had very high worm burdens (Renganathan et al., 1995), and our research activities began with an evaluation of the Zanzibar school-based deworming programme, which started in 1994. Reduction of wasting, stunting, and iron-deficiency anaemia were the primary aims of that programme.

To select the optimal intervention schedule, 12 schools were selected and randomly allocated to non-programme deworming regimens involving treatment twice or three times a year (Stoltzfus et al., 1997a). The anthelminthic treatment was 500 mg generic mebendazole, given as a chewable tablet. Four schools were chosen for twice-yearly treatment, four for treatment three times a year, and four as controls (no treatment); around 1000 children in each group were evaluated at baseline during April–June 1994 and followed up 12 months later. Growth was assessed by height and weight, and iron-deficiency anaemia was assessed on the basis of serum ferritin, erythrocyte protoporphyrin, and haemoglobin levels.

Cross-sectional observations at baseline strongly confirmed our expectation that worm burden predicted nutritional risk when anaemia was the outcome (Figure 2). Hookworm infection intensity, measured as eggs per gram of faeces (epg), was strongly and linearly related to the risk of severe anaemia. A similar strong relationship was observed between hookworm worm burden and low serum ferritin, a measure of iron stores. In a subsample of children we also measured the faecal haem concentration, which is a good indicator of intestinal blood

Figure 1. Initial conceptual framework

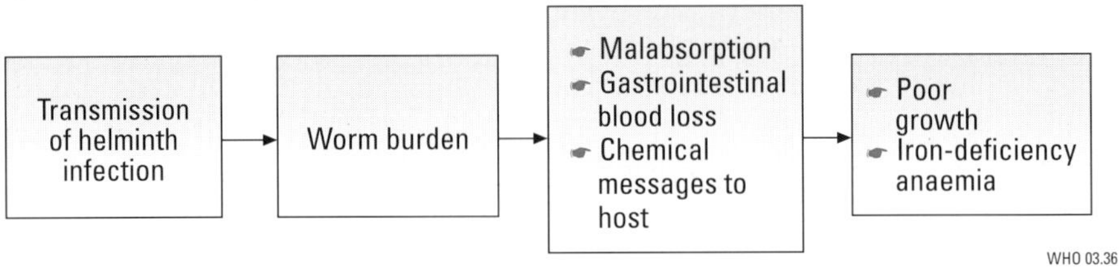

Figure 2. Hookworm infection intensity, faecal haem, and severe anaemia in Zanzibari schoolchildren before intervention[a]

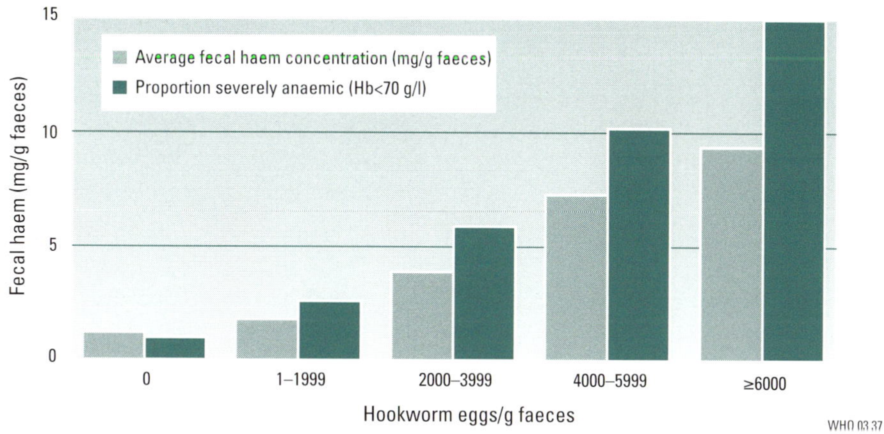

[a] Source of data: Stoltzfus et al., 1996.

loss in people who are not consuming meat (Stoltzfus et al., 1996). There was also a strong linear relationship between intestinal bleeding and hookworm worm burden. These data confirm the long-held view that the number of adult hookworms in the gut directly predicts the quantity of intestinal blood loss and the risk of anaemia in individuals who are not consuming high levels of bioavailable dietary iron (Roche & Layrisse, 1966).

However, when we looked at growth velocities of the children in the placebo group, we found no evidence that smaller weight or height gains were significantly associated with intensity of any geohelminth infection. The only association between helminth infections and growth rates was in children infected with *A. lumbricoides*, who had slightly smaller weight increments than their uninfected peers (1.04 kg/6 months compared with 1.18 kg/6 months, $P = 0.046$), after adjusting for school, age, sex, and stunting at baseline (Stoltzfus et al., 1997b).

The overall effect of the two deworming regimens on iron status was small but significant, and bore a dose–response relationship to the frequency of treatment (see Figure 3). In contrast, the effects on haemoglobin were not significant and not related to frequency of treatment, which suggests that haemoglobin concentration is influenced by important factors other than iron deficiency.

According to the framework in Figure 1, the greatest benefit from anthelminthic treatment should be seen in children with the highest worm burdens. Improvements in iron status were indeed related to the worm burden at baseline, such that the positive effect of deworming on erythrocyte protoporphyrin was greater in children with higher hookworm burdens at baseline (Stoltzfus et al., 1998). The relationship between programme benefit and hookworm burden was also apparent in cases of moderate-to-severe anaemia (see

Figure 3. One-year impact of school-based deworming regimens on iron status outcomes: Zanzibar, 1994–1995[a]

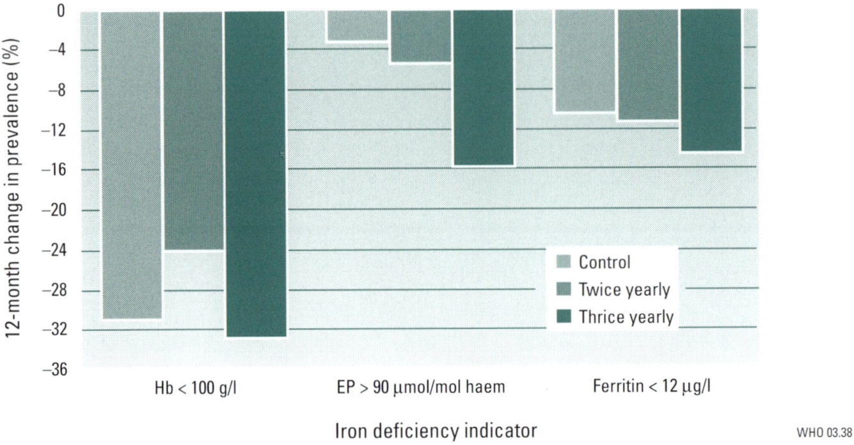

Hb: haemoglobin; EP: erythrocyte protoporphyrin, μmol/mol haem; ferritin: serum ferritin, μg/litre

[a] Source of data: Stoltzfus et al., 1998.

Figure 4): reduction in the incidence of moderate-to-severe anaemia was much greater in children who had higher hookworm burdens at baseline.

When we examined growth outcomes, weight and height gains attributable to the programme regimens were small and were statistically significant only in the subgroup of children who were under 10 years of age at the start of the trial (Stoltzfus et al., 1997b). Programme benefits to growth were greatest (in both absolute terms and in terms of statistical probability) in children who were not stunted at baseline and were under 10 years old; benefits also tended to be greater in boys than in girls. In children under 10 whose height-for-age *Z*-scores were 0 at baseline, the growth benefits attributable to deworming three times a years were 0.45 kg/12 months and 0.5 cm/12 months. However, no relationship between hookworm burden and growth benefit was observed.

To summarize, in relation to Figure 1, what was learned from investigating schoolchildren, the iron deficiency and anaemia results—but not the growth findings—confirmed our expectations. The effects of the anthelminthic programme on growth were difficult to detect overall, and were predicted by factors other than worm burden, namely age, sex, and degree of stunting before intervention.

No clear policies exist for anthelminthic programmes in preschool children, although the nutritional risks are much greater than they are in schoolchildren. Our next study was therefore designed to measure the impact of regular mebendazole treatment on growth and iron-deficiency anaemia in preschool children. Some key preliminary findings, presented here, have led us to revise our thinking about nutritional benefits from anthelminthic treatment.

Figure 4. One-year impact of school-based deworming regimens on moderate–severe anaemia, by hookworm burden: Zanzibar, 1994–1995[a]

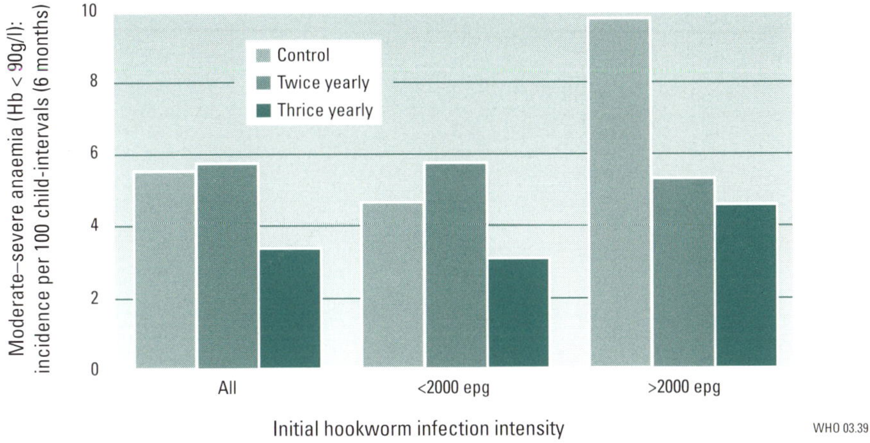

[a] Source of data: Stoltzfus et al., 1998.

The study was a randomized, placebo-controlled trial that employed a factorial design, the factors being daily low-dose iron supplementation, and 3-monthly treatment with 500 mg generic mebendazole. Only the effects of mebendazole are discussed here. From a community census of Kengeja village, 614 children aged 4–71 months entered the trial. The intervention period was 12 months, with baseline and follow-up assessments conducted in September 1996 and 1997. Children were allocated on an individual basis to treatment with mebendazole or placebo. Study staff administered the treatment, and more than 92% of children received at least three of the four planned doses.

The purpose of recruiting children of such a broad age range was to identify more precisely the preschool age at which anthelminthic treatment may be beneficial to nutrition outcomes. Over the first 5 years of life, helminth infection prevalence and intensity increase steeply. On the basis of worm burden, we therefore expected that anthelminthic treatment would be more beneficial to older preschoolers than to the younger children.

As expected, there was a strongly positive relationship between parasitic infections and age (Figure 5): only 5% of children ≥48 months of age were negative for all three helminth infections (Stoltzfus et al., 2000). Haemoglobin concentrations improved with age, and severe anaemia occurred predominantly in children <18 months, in whom 40% of haemoglobin values were <70 g/litre.

In children <30 months of age, hookworm infections were uniformly light (<2000 eggs/g faeces). In cross-sectional relationships at baseline, hookworm infection intensity in children ≥30 months of age was strongly related to both mean haemoglobin and the presence of severe anaemia (Table 1), confirming our

Figure 5. Prevalence of *Plasmodium falciparum* malaria and geohelminth infections by age: Zanzibar, 1996[a]

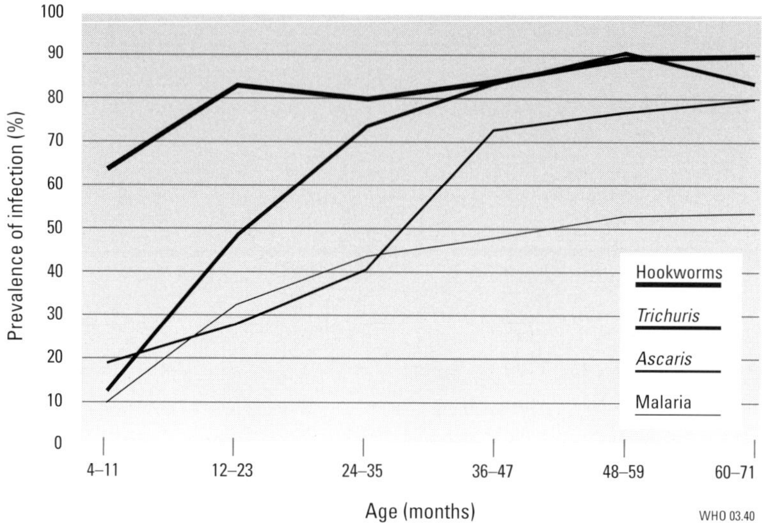

[a] Source of data: Stoltzfus et al., 2000.

Table 1. Haemoglobin and erythrocyte protoporphyrin by hookworm infection intensity in children ≥30 months of age: Zanzibar, 1996[a]

Hookworm infection intensity (eggs/g faeces)	No. of children, *n*	Haemoglobin (g/litre)		Erythrocyte protoporphyrin (μmol/mol haem)
		Mean ± SEM[b]	% <80[c]	Mean ± SEM[c]
0	85	95 ± 1	8.2	141 ± 10
1–1999	156	91 ± 1	12.0	164 ± 8
2000–3999	15	89 ± 3	20.0	148 ± 23
>4000	9	76 ± 4	44.4	222 ± 30
Significance level		$P < 0.001$	$P = 0.005$	$P = 0.044$

[a] Source of data: Stoltzfus et al., 2000.
[b] Least squares mean adjusted for age, recent fever, sex, and malaria parasite density.
[c] Least squares mean adjusted for age, recent fever, and malaria parasite density.

observations in older children. Erythrocyte protoporphyrin also declined with increasing hookworm infection intensity in this older age group, and all of these relationships remained after adjustment for age, sex, recent fever, and malaria parasite density (Stoltzfus et al., 2000).

Mebendazole treatment did not result in significant improvements in haemoglobin or iron status indicators in the group as a whole. Indeed, increments in serum ferritin were much smaller among mebendazole-treated children, although this effect was not statistically significant. However, we do not believe that mebendazole reduced the iron stores of these children; rather, serum ferritin is known to be a positive acute-phase reactant (Elin, Wolff & Finch, 1977), and we speculate that mebendazole reduced the inflammatory response to helminth infection. Although mebendazole did not affect mild anaemia, the incidence of severe anaemia was reduced by about half in the mebendazole-treated group. Examination of this effect by age subgroups showed that the benefit was particularly striking in the youngest children (<24 months), in whom the incidence rates were also the highest. In this subgroup, the reduction in incidence of severe anaemia (defined as haemoglobin <80 g/litre) was more than 50% ($P = 0.10$). This finding, although only marginally statistically significant, was directly contrary to our expectation that benefit would be greatest in children with the highest worm burdens.

Further exploration of mebendazole treatment effects in the older children, however, provided some confirmation for our hypothesis. In children of 4 years and above who were already infected with hookworms at the start of the trial, mebendazole treatment was associated with increased serum ferritin, but in those who were not infected at the start of the trial, mebendazole was associated with reduced serum ferritin (Figure 6). The strength of this interaction was suggestive, but not highly significant. The pattern is consistent with the hypothesis that mebendazole reduced an acute-phase response in children who had not developed a partial immunity to helminth infection and who were exposed to continual transmission; in hookworm-infected children, deworming improved iron stores.

When we examined growth as an outcome, mebendazole had no effect on gains in length. However, wasting malnutrition, defined as weight-for-height Z-score <–2, and the occurrence of small mid-upper-arm circumference (MUAC), was

Figure 6. One-year changes in serum ferritin by mebendazole treatment group and initial hookworm infection in children

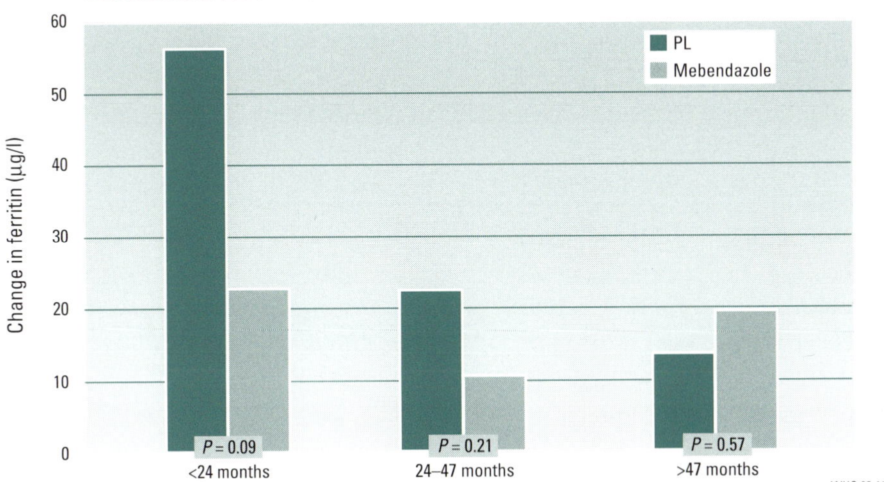

30–40% lower in the treated group. The effect on MUAC <14.0 cm was marginally statistically significant (relative risk 0.68, 95% confidence interval (CI): 0.46–1.00). Further examination of this effect by age group showed that the reduction in wasting malnutrition was limited to children <30 months, and that for this age group the protective effect of deworming was large and significant. The relative risk for MUAC <14.0 cm was 0.59 (95% CI: 0.38–0.92).

This finding is directly contrary to our expectation that the benefits from deworming would be greatest in children with most worms, that is the older preschool children. However, the results are rewarding from a public health point of view, since the burden and consequences of wasting malnutrition are greatest in children under 3 years of age.

These results have led us to revise and expand the framework that guides our research (Figure 7). The worm burden pathway, still included in the upper row of the expanded framework, appears to hold true for iron-deficiency anaemia. But it appears that another pathway exists, which we speculate is mediated by the inflammatory or cytokine response to infection in non-immune individuals. This pathway is depicted in the lower row of the framework, and may have important effects on growth and erythropoiesis in young children in Zanzibar and other environments where helminth infections are acquired early in childhood. The effects on anaemia in this age group cannot be mediated by iron loss, but may rather be mediated through what has been called the anaemia of chronic disease, which may derive in part from cytokine suppression of the bone marrow (Sears, 1992).

Figure 7. Revised conceptual framework, with speculated "inflammatory pathway" for effect of helminth infections on growth and anaemia of chronic disease

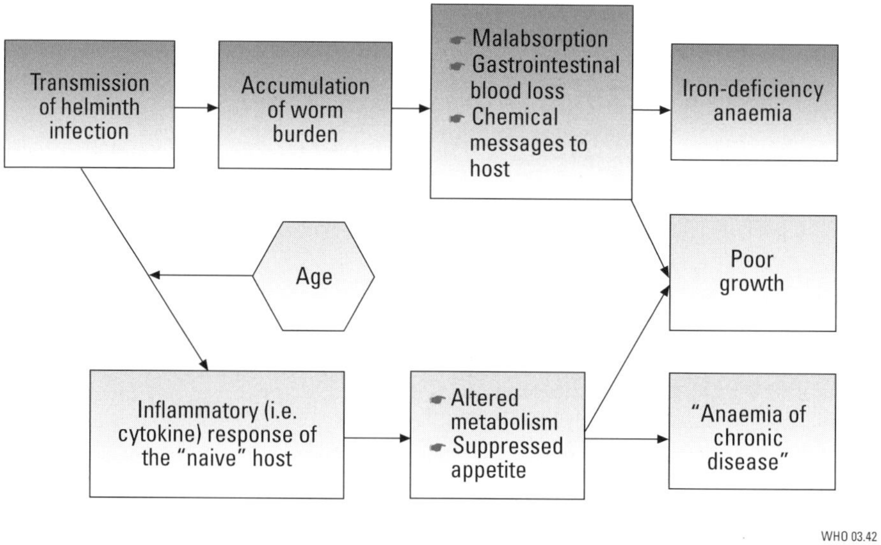

In conclusion, we have summarized results from children in a broad age range that both confirm and expand our prior expectations about the nutritional benefits of anthelminthic treatment. Although this opinion is based on new findings, we propose that geohelminths may cause malnutrition through two distinct pathways. The worm burden pathway predominates in semi-immune individuals, has substantial effects on iron-deficiency anaemia where hookworms are endemic, and—in Zanzibar—has small effects on growth. The inflammation pathway predominates in non-immune individuals, appears to have substantial effects on wasting malnutrition, and may also have substantial effects on the anaemia of chronic disease. The relative importance of these two pathways depends on the age of the target populations, the intensity of transmission, and perhaps the worm species present.

Acknowledgements

We acknowledge the essential contributions of the investigative teams for these two studies. The research team for the school-based deworming evaluation included Kassim Shemel Alawi, Lorenzo Savioli, and James Tielsch. The research team for the preschool study also included Marco Albonico, Jape Khatib Jape, Jane Kvalsvig, Antonio Montresor, Ernesto Pollitt, Lorenzo Savioli, and James Tielsch.

References

Albonico M et al. (1998) Epidemiological evidence for a differential effect of hookworm species *Ancylostoma duodenale* or *Necator americanus* on iron status of children. *International Journal of Epidemiology*, 27:530–537.

Elin RJ, Wolff SM, Finch CA (1977) Effect of induced fever on serum iron and ferritin concentrations in man. *Blood*, 49:147–153.

Renganathan E et al. (1995) Evolution of operational research studies and development of a national control strategy against intestinal helminths in Pemba Island, 1988–92. *Bulletin of the World Health Organization*, 73:183–190.

Roche M, Layrisse M (1966) The nature and causes of "hookworm anemia". *American Journal of Tropical Medicine and Hygiene*, 15:1029–1100.

Savioli L et al. (1990) Control of morbidity due to *Schistosoma haematobium* on Pemba Island: egg excretion and hematuria as indicators of infection. *American Journal of Tropical Medicine and Hygiene*, 43:289–295.

Savioli L, Bundy DAP, Tomkins A (1992) Intestinal parasitic infections: a soluble public health problem. *Transactions of the Royal Society of Tropical Medicine and Hygiene*, 86:353–354.

Sears DA (1992) Anemia of chronic disease. *Medical Clinics of North America*, 76:567–579.

Stoltzfus RJ et al. (1996) Hemoquant determination of hookworm-related blood loss and its role in iron deficiency in African children. *American Journal of Tropical Medicine and Hygiene*, 55:399–404.

Stoltzfus RJ et al. (1997a) School-based deworming programme yields small improvement in growth of Zanzibari school children after one year. *Journal of Nutrition*, 127:2187–2193.

Stoltzfus RJ et al. (1997b) Linear growth retardation in Zanzibari school children. *Journal of Nutrition*, 127:1099–1105.

Stoltzfus RJ et al. (1998) Effects of the Zanzibar school-based deworming programme on iron status of children. *American Journal of Clinical Nutrition*, 68:179–186.

Stoltzfus RJ et al. (2000) Malaria, hookworms and recent fever are related to anemia and iron status indicators in 0- to 5-y old Zanzibari children and these relationships change with age. *Journal of Nutrition*, 130:1724–1733.

Impact of deworming on the growth of schoolchildren in Yangon

Thein Hlaing,[1] Win Win Khine, Tin Tin Than, Myo Khin, Than Saw, San Shwe
Department of Medical Research, Ministry of Health
Yangon, Myanmar

Ma Ma Gyi, Aung Tun, Hkwar Nyo Zin, Phyu Phyu Aye
School Health Division, Department of Health, Ministry of Health
Yangon, Myanmar

Introduction

Intestinal helminth infections and childhood malnutrition generally occur in the same people at the same time in the same parts of the world (Crompton, 1986). This is certainly true of Myanmar, where a survey among schoolchildren in Yangon in 1968 indicated that the prevalence of *Ascaris lumbricoides* and *Trichuris trichiura* infection was over 80% (Hla Myint, 1970), while a nationwide nutrition survey showed that the prevalence of school-age malnutrition is over 40% (National Nutrition Centre, 1991). Similar prevalences of helminth infection and malnutrition were found in rural Myanmar (Hlaing et al., 1991).

A number of studies have shown that children's growth can be improved by mass treatment with broad-spectrum anthelminthic drugs in populations with a high prevalence of helminth infections (Willett, Kilama & Kihamia, 1979; Stephenson et al., 1989; Hlaing et al., 1991; Stoltzfus et al., 1997). Mass treatment with piperazine has been provided to primary-school children throughout Myanmar on an annual basis since the early 1980s. Piperazine is a drug with low efficacy against helminth infections (Gupta & Urrutia, 1982), so the impact of this nationwide treatment on the nutritional status of schoolchildren has not been evaluated.

In 1992, a supply of levamisole became available to the School Heath Division of the Department of Health in Yangon, and an operational study was undertaken to assess the impact of a programme of deworming primary-school children every 6 months during 1993, 1994, and 1995. Such a study was deemed important to provide information to policy-makers and health managers on the potential value of the deworming programme.

[1] Dr Thein Hlaing, Trauma Services, Parkview Hospital, 2200 Randallia Road, Fort Wayne, IN 46805, USA (present address)
Tel: +1 219 484 6636, Fax: +1 219 480 7146, E-mail: thein.hlaning@parkview.com

Materials and methods

Study area

Two suburban townships, one in the east (Tharkeyta, or Area 1) and one in the north (Mingaladon, or Area 2) of Yangon, were chosen as the study areas because they had a comparatively low socioeconomic status among the 32 townships within the city, they were separated by one or more townships of similar socioeconomic status, and township and ward administrative bodies were cooperative. Area 1 was a new settlement in the early 1960s, smaller, and more populated than Area 2. However, environmental sanitation conditions are similar in the two townships. Before the study began in each area, personnel from the School Health Division of the Department of Health and from the Department of Medical Research, Ministry of Health, held discussions with the township administration and education authorities, ward elders, and the children's parents. The purpose and benefits of deworming and the need for full cooperation for the success of the project were carefully explained.

Six of the 46 primary schools in Area 1 and seven of the 35 in Area 2 were randomly selected. Primary schools in Myanmar have five classes: kindergarten and grades 1–4. The academic year begins in June and ends in March the following year. There were two school health teams, one in each study area. Normally, the health and nutrition of schoolchildren are the responsibility of the school health team in the township concerned; the team comprises a medical officer, a child officer, a clerk, and a nurse. The study protocol was reviewed by the Ethical Committee at the Department of Medical Research and was approved by the Committee and the responsible personnel of the Department of Health. In each township, the field research team in each area comprised the school health team members plus a research officer from the Department of Medical Research, working in cooperation with schoolteachers and parents.

Study subjects

The study subjects included 1800 schoolchildren in Area 1 and 1525 in Area 2—a total of 3325 pupils from kindergarten to second grade. Because of late enrolment in schools in the area, pupils' ages ranged from 5 to 13 years. The age of each child was obtained from the school records, which are taken from birth certificates—a requirement for school registration. When school records of age were missing, parents were consulted. At the baseline, i.e. in December 1993, each child's height was accurately measured (to the nearest 0.1 cm) by the trained school health nurse and the clerk in each area. Similarly, the medical officer and the research officer weighed each child accurately (to the nearest 0.1 kg). Subsequently, all the children in the schools (including those in grades 3 and 4) were treated with 80 mg (or 2 tablets) of levamisole. A total of 796 stool samples were taken from all the first-grade pupils (485 in Area 1 and 311 in Area 2); these were collected in plastic bags after appropriate instruction had been given to the parents or guardians. Three slides for each stool sample were examined for intestinal parasites by direct faecal smear; examinations were carried out in the Parasitology Research Division of the Department of Medical Research.

The second and third anthropometric measurements of children from kindergarten to second grade were taken in June 1994 and December 1994 by the same personnel that performed the first measurements. Following each anthropometric measurement, levamisole was again administered to all children in the schools (those included and those not included in the study), in the same manner as at baseline.

Analysis

At 6 months, measurements were taken from 1546 pupils (85.9% of the original sample) in Area 1 and from 1318 (86.5%) in Area 2. The corresponding figures after 12 months were 1358 (75.4%) and 1233 (80.9%) respectively.

Children aged 10 years and over were excluded from the analysis. Data from the remaining 2439 (76.3%) schoolchildren aged 5–9 years who had received anthelminthic treatment at 0, 6, and 12 months and for whom there were complete sets of height and weight measurements were analysed for any changes in growth and nutritional status. National Center for Health Statistics (NCHS) growth charts were used to grade nutritional status according to height-for-age and weight-for-age (WHO, 1983). Chi-square tests were applied to evaluate the differences between proportions in the two samples. Student *t*-tests were carried out to show the significant differences between two samples (independent sample *t*-tests) or within the same sample at different times of follow up (paired *t*-tests).

Results

Prevalence of major intestinal parasites

The prevalence rates of *Ascaris lumbricoides*, *Trichuris trichiura*, and *Giardia lamblia* among schoolchildren in Area 1 were much higher than those among children in Area 2 (Table 1). For *A. lumbricoides*, and *T. trichiura* these differences were very highly significant ($P < 0.001$).

Initial nutritional status

In general, the schoolchildren in study Area 1 were slightly taller and heavier than those in Area 2 by age. However, the differences were significant only for height at age 6 years in males (Figure 1a) and at ages 6–8 years in females (Figure 1b). The mean weight-for-age values in males and females were similar (Figures 1c and 1d). Weight differences between the areas were significant in females at ages 5 and 6 years.

Only moderate and severe malnutrition were taken into consideration in the results. The prevalence rates of moderate and severe malnutrition among the children in both areas were high, ranging from 45 to 65%. The rates were higher in

Table 1. Prevalence of major intestinal parasites by study area at baseline

Intestinal parasite	Prevalence[a] (%)		*P*-value
	Area 1 ($n = 479$)	Area 2 ($n = 309$)	
A. lumbricoides	59.8 (46.8–81.3)	36.6 (25.8–44.1)	P < 0.001
T. trichiura	32.1 (16.2–47.9)	11.0 (6.5–20.5)	P < 0.001
G. lamblia	4.4 (1.4–8.3)	2.6 (0–6.7)	P < 0.05

[a] Range of prevalence in parentheses.

Table 2. Baseline level of malnutrition by study area in both sexes

Age (years)	Numbers of malnourished children based on height-for-age				Numbers of malnourished children based on weight-for-age			
	Area 1		Area 2		Area 1		Area 2	
	n	%	*n*	%	*n*	%	*n*	%
5	249	46.2	169	59.2[a]	218	52.8	165	58.8
6	329	50.2	263	63.9[b]	303	58.7	255	66.7
7	349	44.7	276	56.5[a]	334	47.9	278	48.6
8	163	65.6	255	63.5	162	60.5	252	55.6
9	102	58.8	119	66.4	100	60.0	119	58.8

[a] Significantly different from Area 1 ($P < 0.01$).
[b] Significantly different from Area 1 ($P < 0.001$).

Area 2 than in Area 1 (Table 2). From 5 to 7 years of age the differences were highly significant when classified by height-for-age but not by weight-for-age. Findings by sex in each area were similar.

Height and weight gain by study area

After two treatments with levamisole, 6 months apart, increments of height and weight were greater among children in study Area 1 than among those in Area 2 (Figure 2): the height gain was about 6 cm in Area 1 and 5 cm in Area 2 in all ages and sexes. The differences in height gains by age between the two areas were highly significant from 5 to 8 years in both sexes (Figure 2, a and b). The weight gain was 3–4 kg in both males and females in Area 2, but in Area 1 it was 4–5.5 kg in males and 4–6.5 kg in females, with increasing trend, from 6 to 10 years (Figure 2, c and d). The differences in weight gain in the two areas were significant to very highly significant in both sexes from 6 to 10 years. The exception was females aged 5 years, for whom weight gain was significantly higher in Area 2 than in Area 1.

Figure 1. Mean height and weight of schoolchildren by age in Area 1 and Area 2

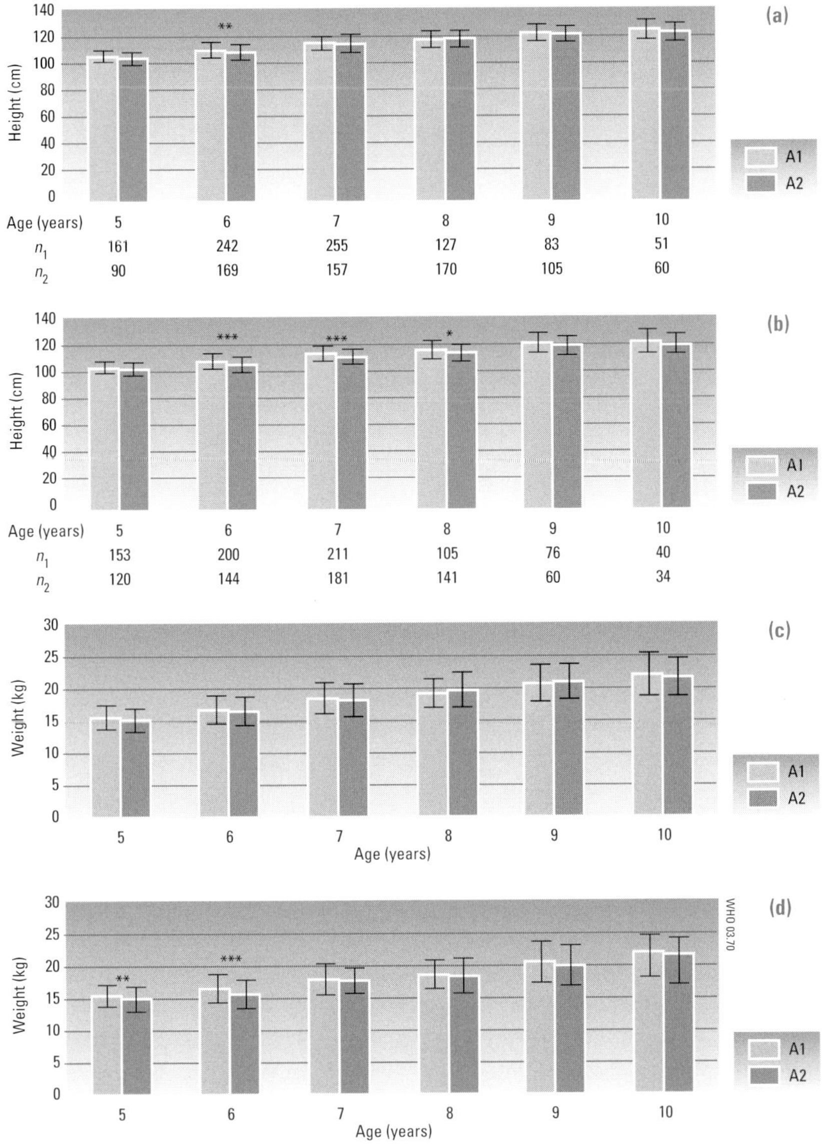

(a) Mean height ± SD of males
(b) Mean height ± SD of females
(c) Mean weight ± SD of males
(d) Mean weight ± SD of females

Notes:

1. Statistical significance of differences is indicated as follows:
   * $P < 0.05$
   ** $P < 0.01$
   *** $P < 0.001$
   In all other cases, differences are not significant ($P > 0.05$).
2. Samples sizes in Figures 1(a) and 1(c), and in 1(b) and 1(d) are the same.

Figure 2. Mean height gain and weight gain of schoolchildren by age in Area 1 and Area 2

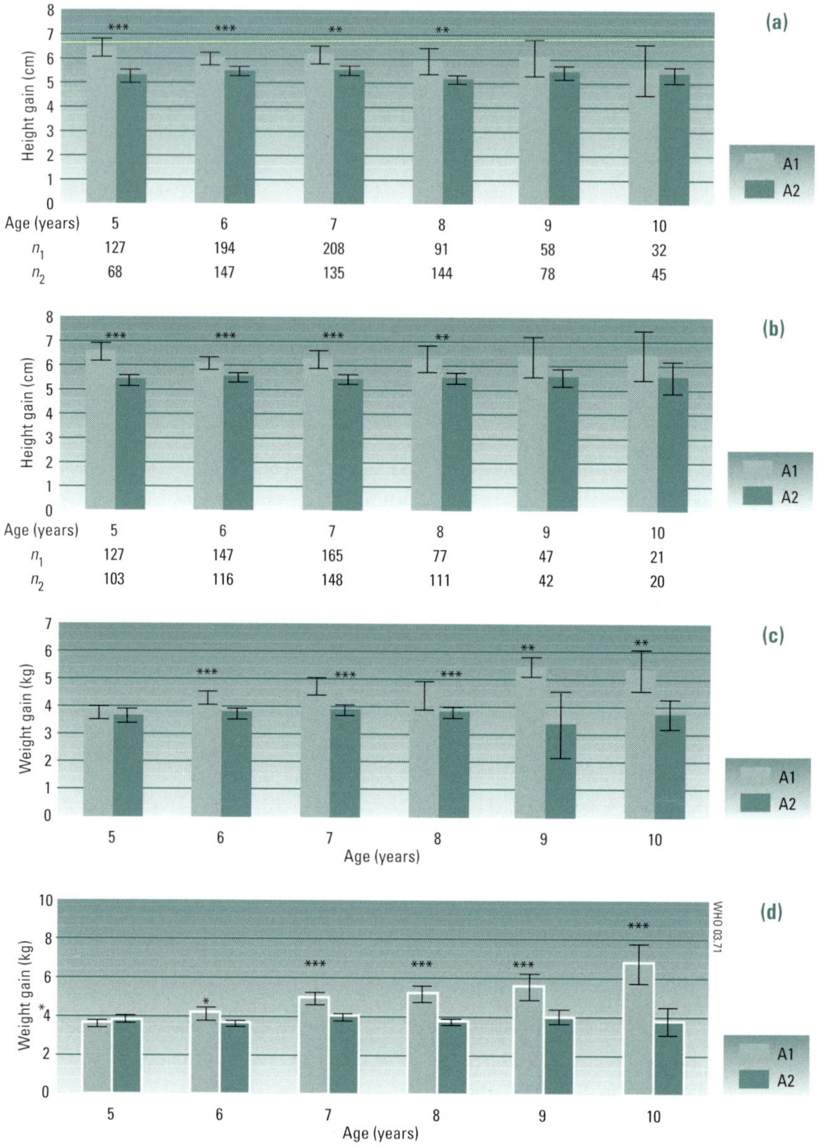

(a) Mean height gain ± 1.96 SEM of males
(b) Mean height gain ± 1.96 SEM of females
(c) Mean weight gain ± 1.96 SEM of males
(d) Mean weight gain ± 1.96 SEM of females

Notes:

1. Statistical significance of differences is indicated as follows:
   * $P < 0.05$
   ** $P < 0.01$
   *** $P < 0.001$
   In all other cases, differences are not significant ($P > 0.05$).
2. Samples sizes in Figures 2(a) and 2(c), and in 2(b) and 2(d) are the same.

Height and weight gain by time period

The gain in mean height in male (Figure 3a) and female (Figure 3b) schoolchildren was more or less equal—about 2.5–3.0 cm in the first and second 6-month periods across the ages—and the difference by period was significant only at 6 years in males and 7 years in females. The gain in weight in males (Figure 3c) and females (Figure 3d) was about 1 kg in the first period and about 3 kg in the second in every age group, and the differences between the periods were very highly significant in all ages and sexes.

Reduction in malnutrition before and after deworming

The degree of malnutrition exhibited by the children was graded using WHO criteria for nutritional status according to weight-for-age or height-for-age, and the results after 6 to 12 months were compared with appropriate (for age) pretreatment baseline levels. Malnutrition declined (by 3–8%) across the age groups when classified according to height-for-age. The reduction was significant in males only at age 7 years and in females at age 5 (Table 3). However, there were marked reductions (30–50%) in the severity of malnutrition when classified according to weight-for-age, and these were very highly significant in all ages and sexes (Table 4). For females aged 6–9 years the reduction was in decreasing order.

Table 3. Effects of 6-monthly deworming treatment on malnutrition by sex and age as assessed by height-for-age

Age (years)	No. of malnourished children	Malnutrition		Reduction in malnutrition	
		Baseline	After treatment	Before and after treatment	With and without treatment[a]
Males					
5	189	53.9	49.2	4.8	
6	335	58.2	54.9	3.3	9.0[b]
7	326	51.8	43.9	7.9[b]	−3.1
8	232	66.8	59.1	7.7	22.9[c]
9	134	62.7	55.2	7.5	3.6
Females					
5	229	49.3	41.9	7.4[a]	
6	257	53.7	48.2	5.5	11.8[c]
7	299	47.8	41.5	6.3	−0.4
8	186	61.3	58.1	3.3	19.8[d]
9	87	63.2	65.5	−2.3	5.1

[a] Reduction in malnutrition is estimated by subtracting malnutrition in treated children from that in untreated children 1 year older, e.g. malnutrition at 6 years for males = 58.2 − 49.2 = 9.0.
[b] $P < 0.05$.
[c] $P < 0.01$.
[d] $P < 0.001$.

Figure 3. Mean height and weight gain of schoolchildren by age in first and second 6-month periods

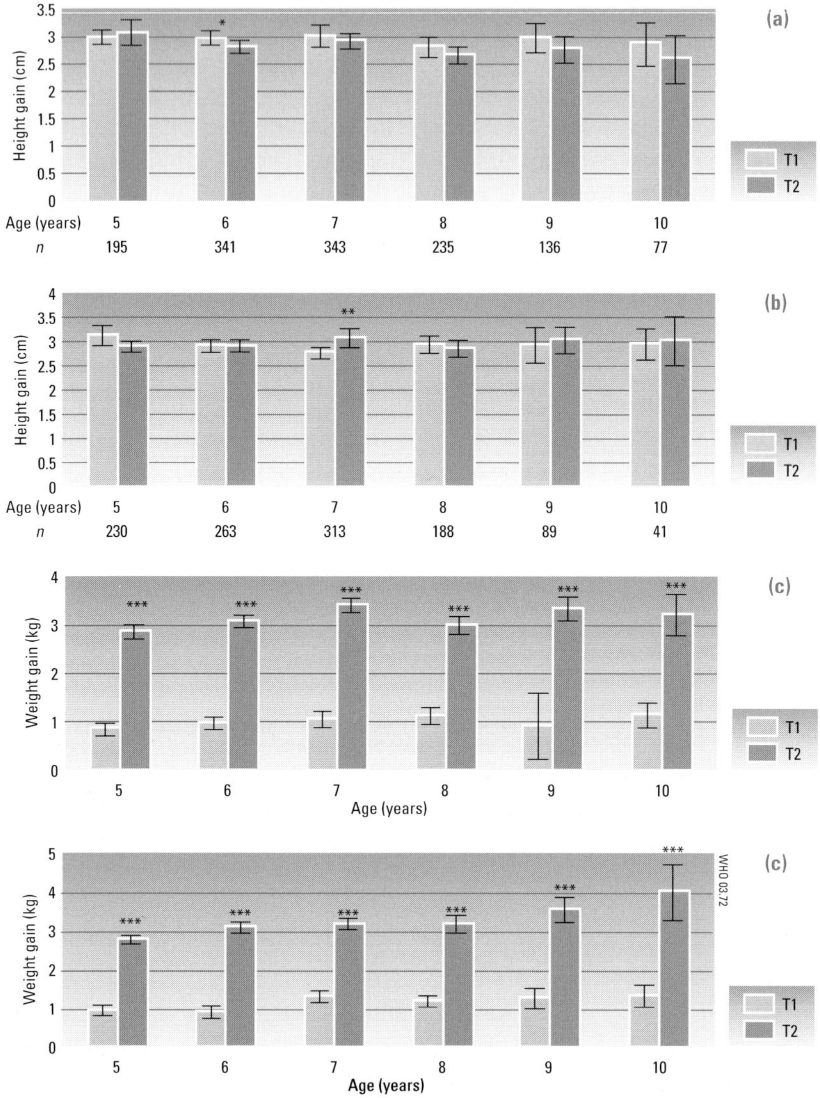

(a) Mean height gain ± 1.96 SEM of males in first 6-month period
(b) Mean height gain ± 1.96 SEM of females in first 6-month period
(c) Mean weight gain ± 1.96 SEM of males in second 6-month period
(d) Mean weight gain ± 1.96 SEM of females in second six-month period

Notes:

1. Statistical significance of differences is indicated as follows:
   * $P < 0.05$
   ** $P < 0.01$
   *** $P < 0.001$
   In all other cases, differences are not significant ($P > 0.05$).
2. Samples sizes in Figures 3(a) and 3(c), and in 3(b) and 3(d) are the same.

Table 4. Effects of 6-monthly deworming treatment on malnutrition by sex and age as assessed by weight-for-age

Age (years)	No. of malnourished children	Malnutrition		Reduction in malnutrition	
		Baseline	After treatment	Before and after treatment	With and without treatment[a]
Males					
5	177	63.8	17.5	46.3[b]	
6	322	67.7	19.6	48.1[b]	50.2[b]
7	314	55.7	15.3	40.4[b]	36.1[b]
8	232	57.3	25.0	32.3[b]	42.0[b]
9	132	60.6	19.7	40.9[b]	36.5[b]
Females					
5	206	48.1	6.3	41.7[b]	
6	236	55.1	7.6	47.5[b]	48.8[b]
7	298	40.3	6.0	34.3[b]	32.7[b]
8	182	57.7	24.7	33.0[b]	51.7[b]
9	87	57.5	27.6	29.9[b]	32.8[b]

[a] Reduction in malnutrition is estimated by subtracting malnutrition in treated children from that in untreated children 1 year older, e.g. malnutrition at 6 years for males = 67.7 – 17.5 = 50.2.
[b] $P < 0.001$

Reduction in malnutrition with and without deworming

The prevalence of malnutrition at the end of the study was also compared with the prevalence that was detected among the children who had the same age at baseline, i.e. were not yet treated. Thus children aged 7 years at the baseline were 8 years old by the time of anthropometric measurements at month 12 and had received two treatments (at the baseline and at month 6); they were compared with untreated (baseline) children aged 8 years.

Before- and after-treatment comparisons showed that the reductions in malnutrition were in the range 0–23% in males and 0–20% in females across the ages when classified by height-for-age (Table 3). The reductions were significant at 6 and 8 years in both males and females. However, there were greater reductions (30–50%) in the severity of malnutrition based on weight-for-age, and these were very highly significant (Table 4).

Discussion

The greater prevalence of the major intestinal parasites in Area 1 than in Area 2 may be due to the relatively greater population density or to crowded housing conditions in Area 1. Conversely, the slightly higher nutritional status in Area 1 than in Area 2 may be the result of better incomes. Several investigators have reported that growth deficits in children can be attributed to *A. lumbricoides* (Willett, Kilama & Kihamia, 1979; Stephenson et al., 1989; Hlaing et al., 1991),

T. trichiura (Cooper & Bundy, 1986), and *G. lamblia* (Gupta & Urrutia, 1982) infections. The greater height or weight gains among schoolchildren in Area 1 than in Area 2 after deworming may be the result of elimination of the more severe infections leading to better appetite and food intake. Place may therefore play an important role in affecting the growth in schoolchildren after deworming.

The growth increment in terms of height and weight following deworming was not the same in the two 6-month periods of the study. Although the height gain was similar during the two periods, weight increases were markedly greater during the second period. The unevenness of weight gain in the two periods might be due to subsequent reduction of worm intensity in the children after repeated drug treatment (Hlaing, Saw & Lwin, 1987). The consistent pattern of improvement in height and weight by age and sex (Figure 2) illustrates the effect of deworming on growth over time.

Reductions of 30% of more in moderate and severe malnutrition (when classified by weight-for-age) among schoolchildren as the result of the treatment highlights the remarkable impact of deworming on the growth and health of schoolchildren in Myanmar.

Since this was an operational type of study, we employed a comparatively simple study design and did not use a separate control group of schoolchildren. However, we have compared the growth increment in treated and untreated children in similar age groups by looking at anthropometric measurements taken over a period of one year. The factors influencing the nutritional status of the children were unlikely to change during this relatively short period. A study design without a separate control group minimizes both the cost of programme evaluation and the time needed.

In this study, weight-for-age appeared to be better than height-for-age as an indicator of the changes in nutritional status that resulted from deworming.

Summary

A one-year operational research study was started in December 1993 to investigate the impact of 6-monthly treatment of parasitic infections with levamisole on the growth of primary-school children in two townships (Area 1 and Area 2) of suburban Yangon. Measurements of height (to the nearest 0.1 cm) and weight (to 0.1 kg) were recorded for 3325 primary-school children (from kindergarten to second grade) at 0, 6, and 12 months; drug treatment followed anthropometric measurements at 0 and 6 months. At the baseline, 796 stool samples were collected from all the first-grade pupils in the study and examined for intestinal parasites.

The prevalence rates of *Ascaris lumbricoides*, *Trichuris trichiura* and *Giardia lamblia* among schoolchildren in Area 1 (58.9, 32.1, and 4.4% respectively) were much higher than those among children in Area 2 (36.6, 11.0, and 2.6% respectively). After 1 year of deworming, height and weight gains were greater in Area 1 than in Area 2. Height gain in both males and females was about the same (2.5–3 cm) in the first and second 6-month periods across the ages; weight gain in males or females was three times higher in the second (3 kg) than in the first period (1 kg) by age. The two 6-monthly drug treatment rounds reduced moderate and severe malnutrition by 30% or more when malnutrition was classified by weight-for-age. Our study—the first of its kind in Myanmar—has thus fulfilled its purpose; the health policy implications of its results are such that it can help to improve school health services and ultimately the nutrition and general health of schoolchildren in the country.

Acknowledgements

We wish to thank the township administrative bodies, town elders, and school principals and teachers in Thakeyta and Mingaladon townships for their cooperation in this project. Thanks are also due to the staff of the Epidemiology and Parasitology Research Divisions of the Department of Medical Research, and School Health Team members of the Department of Health. The principal author is particularly grateful to the Director of Trauma Services, Parkview Hospital, Fort Wayne, USA, for kindly allowing him the use of facilities for further data analysis and preparation of this paper. The project was partly funded by a research grant from the Department of Medical Research in Yangon.

References

Cooper ES, Bundy DAP (1986) Trichuriasis in St Lucia. In: McNeish AS, Walker-Smith JA, eds. *Diarrhoea and malnutrition in children*. London, Butterworths:91–96.

Crompton DWT (1986) Nutritional aspects of infection. *Transactions of the Royal Society of Tropical Medicine and Hygiene*, 80:697–705.

Gupta MC, Urrutia JJ (1982) Effect of periodic anti-*Ascaris* and anti-*Giardia* treatment upon nutritional status of preschool children. *American Journal of Clinical Nutrition*, 36:79–86.

Hla M (1970) Intestinal parasitism in primary school children in Greater Rangoon. *Union of Burma Journal of Life Sciences*, 3:141–145.

Hlaing T, Saw T, Lwin M (1987) Reinfection of people with *Ascaris lumbricoides* following single, 6-month and 12-month interval mass chemotherapy in Okpo village, rural Burma. *Transactions of the Royal Society of Tropical Medicine and Hygiene*, 81:140–146.

Hlaing T et al. (1991) A controlled chemotherapeutic intervention trial on the relationship between *Ascaris lumbricoides* infection and malnutrition in children. *Transactions of the Royal Society of Tropical Medicine and Hygiene*, 85:523–528.

National Nutrition Centre (1991) *Nutrition situation of Myanmar children: preliminary report of the nutrition surveys*. Yangon, Department of Health.

Stephenson LS et al. (1989) Treatment with a single dose of albendazole improves growth of Kenyan school children with hookworm, *Trichuris trichiura* and *Ascaris lumbricoides* infections. *American Journal of Tropical Medicine and Hygiene*, 41:78–87.

Stoltzfus RJ et al. (1997) School-based deworming program yields small improvement in growth of Zanzibari school children after one year. *Journal of Nutrition*, 127:2187–2193.

WHO (1983) *Measuring change in nutritional status: guidelines for assessing the nutritional impact of supplementary feeding programmes for vulnerable groups.* Geneva, World Health Organization.

Willett WC, Kilama WL, Kihamia CM (1979) *Ascaris* and growth rates: a randomized trial of treatment. *American Journal of Public Health*, 69:987–991.

Parasites, nutrition, child development, and public policy

J.D. Kvalsvig[1]
Child Development Programme
School of Anthropology and Psychology
University of Natal, Durban, South Africa

Introduction

Most public policy with respect to intestinal parasites is built on the premise that children would be better off without them. The evidence in the research literature for a causal link between parasitic infection and impaired cognitive function or delayed cognitive development is by no means complete. Review articles in the past few years have returned cautious verdicts regarding the beneficial effect of treatment: "not proven" but "plausible" (Connolly & Kvalsvig, 1993; Connolly, 1998; Dickson et al., 1999), "persuasive" (Nokes & Bundy, 1994), or "a qualified yes" (Watkins & Pollitt, 1997). These reviews have raised a number of questions about the way in which research in this field has been conducted and the conclusions that may legitimately be drawn from this body of work. This paper takes a developmental view of the questions raised in the reviews, and outlines some of the implications for public policy.

The long view

Developmental psychology is naturally concerned with changes over time in the behavioural and cognitive functioning of children and adults. We speak of developmental *trajectories* (Wachs, 1996) to refer to increments over time in a particular developmental functional domain. The *transactional* nature of development refers to "the interplay between child and context across time" in what Sameroff (1987) terms a "continuous dynamic process"; *transitional periods* refer to periods of rapid qualitative change in behaviour and cognition, such as adolescence or the time of entry into school, which are thought to be times when negative contexts might have a more permanent effect.

Breaking with this alliterative terminology, we might also think of *stages* in cognitive development, with each stage in the construction of mind built on preceding stages. Figure 1 shows this diagrammatically, with four major domains of developmental change before the age of 10 years—motor, language, sociability,

[1] Dr J.D. Kvalsvig, Child, Youth and Family Development, Human Sciences Research Council, 359 King George V Avenue, Durban 4001, Natal, South Africa
Tel: +27 31 273 1429, Fax: +27 31 273 1403, E-mail: jkvalsvig@hsrc.ac.za

Figure 1. The developmental sequence and the increasing prevalence of parasite infections in a sample of preschool children in Zanzibar

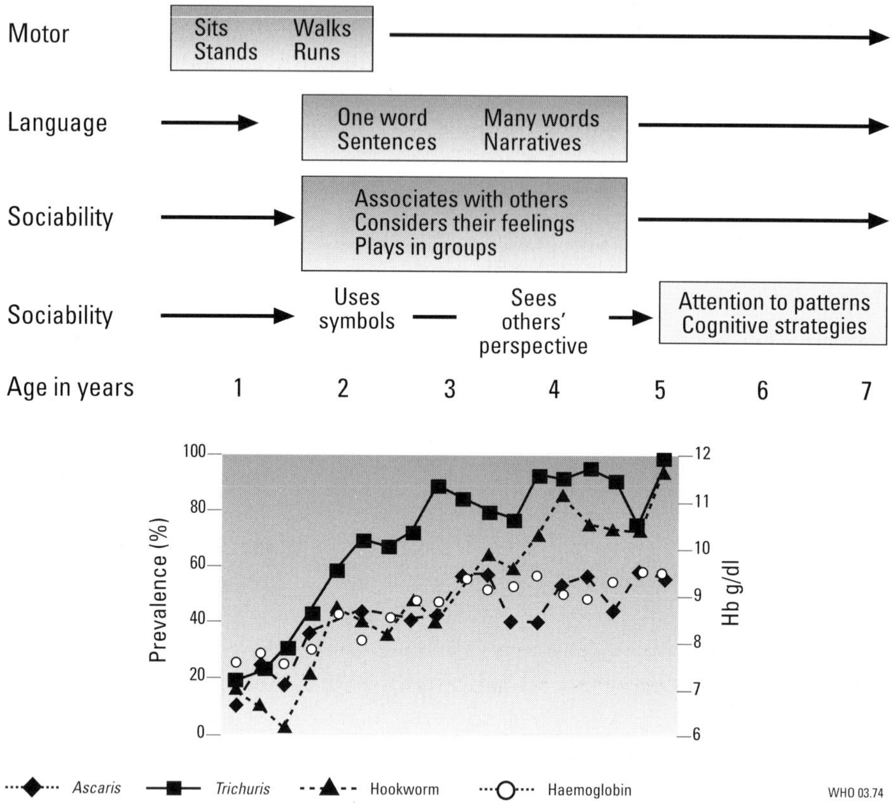

and cognition. These domains are set against graphs depicting the increase in prevalence of three species of parasites in Zanzibari children. The graph also depicts the increase in haemoglobin with age over the same period.

Functional development does not proceed at a steady pace in any of the domains depicted in Figure 1. In the domain of language development, for example, babbling is a precursor to language; the development of skills continues over the age range, but there is a period of rapid acquisition when changes seem to occur on a daily basis. Once a child has achieved a reasonable level of competence, further development becomes more subtle and is strongly aligned to cognitive development.

Figure 1 illustrates how these developmental spurts occur at different times and how competence in one domain can have an important influence on the later development of competence in another. Thus, learning to use words is in effect learning to use symbols, a valuable cognitive skill. Similarly, as children become more sociable, they learn to see things from another's point of view—an early stage in developing a theory of mind and working with abstract ideas. In a Piagetian model, cognitive development is hierarchical, with new stages resting on previous experience. Any disruptions to these processes caused by nutritional deficits and parasitic infections have the capacity to disturb the acquisition not only of current thinking skills but also of future thinking skills.

At the same time as these observable and measurable changes are taking place, the underlying development of neural connections in the central nervous system is proceeding in stages as myelination allows for new areas of the brain to become fully functional. The language areas become active at about 18 months, the reticular formation, which is involved in maintaining attention, is not fully myelinated until puberty, and the frontal lobes, implicated in thinking and planning, continue myelination into adulthood. The functional organization of the brain is more modifiable early on in development and becomes more rigid later (Carter, 1998).

Nutritional deficits and common parasite infections have their own age-related rhythms. Nutritional deficits are thought to have more severe consequences in younger children, and supplementation is more successful before the age of 2 years (Pollitt, 1996). Geohelminth infections are usually acquired at ages when children start to move around independently and play on the ground outdoors. The acquisition of new infections probably overlaps with the time of weaning, exacerbating deficiencies if the supply of nutrients is already vulnerable. The trend lines in Figure 1 for parasites and haemoglobin are derived from measurements in a village in Zanzibar where malaria is holoendemic. Haemoglobin levels are very low across all the age groups, although they do rise steadily in an age-related manner. It is thought that these depressed haemoglobin levels are attributable to malarial infections in infants and very young children, and to parasitic infections at a later stage, demonstrating a complex relationship between age and nutritional risk (Stoltzfus, 1999, personal communication).

The risk and resilience models of child development have demonstrated the multifactorial nature of poverty: single traumatic incidents or potentially deleterious factors do not inevitably have detectable consequences, but the convergence of several harmful factors is more likely to show an effect. Even in regions of exceptional poverty there are children who appear to develop quite normally emotionally and intellectually, seemingly protected by fortunate circumstances (Werner, 1989). Following this line of reasoning, it is unlikely that deworming will benefit all children equally or similarly. As with any other single intervention, deworming will be one of many influences on the course of development. Its impact will depend on the number and salience of environmental constraints and enabling factors, and the developmental stage of the child.

This is the framework in which we must view and assess the evidence for the impact of parasitic infections on cognition, and the consequent public health policy decisions.

Questions raised by recent reviews

The review papers referred to above are in reasonable agreement over the important issues in this field.

Design

Dickson et al. (1999) reviewed reports of 27 randomized or quasi-randomized trials in which drug treatment was compared with placebo or with no drug treatment. The authors concluded that "while the treatment may have some positive effects on children's growth there was not enough evidence to evaluate the effect of anthelminthic drugs on cognitive performance". With regard to the success or otherwise of implementing this design, the authors note that it is difficult to ensure that the treatment allocation is concealed. There is a further point: in trials extending over many months, it may also be difficult to ensure that children are not treated independently by worried parents if low-cost, over-the-counter remedies are available.

As researchers try to address the inadequacies in earlier studies, there is a trend towards better-designed studies but no improvement in the consistency of the results. In fact, the older studies show harmful effects of parasites more consistently than more recent work (Watkins & Pollitt, 1997). The consensus from reviews seems to be that the current lack of consistent results is not necessarily, or even probably, an indication that treatment has no value. The environmental contexts being studied are very variable; within those, the prevalence and intensity of the infections vary greatly. Moreover, these newer, better-designed studies are few in number.

As evidence accumulates for the health and nutritional benefits of deworming, and the safety, effectiveness, and affordability of targeted treatment programmes become more apparent (Albonico et al., 1999), ethical considerations may prevent approval from being granted for randomized placebo-controlled trials. In defence of the continuation of such trials it may be argued that the long-term effects of treatment are not yet known, but ethical concerns need careful consideration in the design of new trials.

Species and intensity

The question of assessing the impact of geohelminth infections on child development can be approached from a biological perspective. Knowing more about the health and nutritional impact of these parasites would obviously help to narrow the choice of cognitive and behavioural assessment measures.

Three parasite species—*Ascaris lumbricoides*, *Trichuri trichiura*, and *Necator americanus*—are commonly targeted by control programmes. They are dissimilar in many respects and likely to affect children differently, as is frequently emphasized in published papers; moreover, different combinations of species may elicit different host responses. The life cycles and modes of transmission of the three species are different. For example, hookworms have free-living larvae, which penetrate the skin of the human host, whereas the eggs of the other two species are ingested. Feeding methods are different: hookworms feed directly from human tissue, causing blood loss, but *A. lumbricoides* and *T. trichiura* compete with their human hosts for food before it is absorbed. Adult *A. lumbricoides* are bigger than other worms and known to cause intestinal obstructions. Taking all these factors into consideration, it is probable that the species differ in the amount and type of nutritional loss that they cause and the degree of discomfort and damage suffered by the host.

Nevertheless, there may also be commonalities in the type of damage inflicted and the host response. It is likely that there are energy costs to the host in harbouring parasites that are so directly linked to the absorption of nutrients, and there may be a generalized immune response in the presence of invader organisms that is manifest as lethargy or passivity in the infected child.

A few examples of the differences in characteristic health problems associated with each of the three species should suffice to make the point that the particular species endemic to an area may be important in determining the nature of the cognitive impairment that children experience.

All three species have been associated with poor iron status in Zanzibari schoolchildren (Stoltzfus et al., 1998), but the association with hookworms was the strongest. The negative association between iron deficiency, development, and educational performance is well documented, and there is reason to believe that, in hookworm-endemic areas, intestinal parasites are an important constraint in the cognitive development of the children. Supporting evidence for this comes from a recent study in Indonesia (Sakti et al., 1999), where hookworm was significantly associated with poorer performance on cognitive tests at baseline, after controlling for possible confounders in the analysis. In the light of the transactional model of development outlined above, it is interesting that this association was more pronounced in older children.

Ascariasis was linked with poor lactose digestion in a study in Panama (Taren et al., 1987), and the large worms are also a cause of intestinal obstructions, particularly in children. This implies a different mechanism by which this species may affect children's nutritional status and possibly their educational performance. *T. trichiura* is sometimes associated with diarrhoea and rectal prolapse, implying inflammatory responses and yet another set of deleterious factors (Callender et al., 1993).

Until common and differential effects are distinguished, we may not know with any precision where deworming priorities lie. The situation is further complicated by the fact that most treatments of choice are more effective against *A. lumbricoides* than against *T. trichiura*, so the relative prevalence of species changes as deworming programmes proceed. Once a treatment programme has all but eliminated *A. lumbricoides* from a community, programme administrators need to know the implications for the well-being of children of the residual *T. trichiura* infections. This issue is of particular relevance for public health programmes.

A related question deals with intensity of infection. Is it possible to identify threshold intensities beyond which growth or cognitive development is compromised? This issue developed a strange twist with the report from Pollitt et al. (1991) that, although the educational performance of severely infected children improved following treatment, that of lightly infected children actually deteriorated. Watkins & Pollitt (1997) speculated that parasites may have some beneficial effects or, alternatively, that treatment may do some harm. The process of ascertaining the severity of an infection in order to target only severely infected children could add significantly to the cost of a treatment programme. Dickson et al. (1999) propose to reanalyse data from the trials included in the Cochrane database to assess intensity effects more thoroughly; for the present, however, on the basis of a few studies that examined this effect, it is assumed that benefits are proportional to the intensity of infections.

What aspects of behaviour and psychological function are affected?

Before the developmental sequence is considered in detail, it is instructive to look at the types of behavioural cognitive measures that have been used or suggested for use.

Dickson et al. (1999) comment that, because the biological causal mechanisms are unknown, it was difficult to know which aspects of cognition might be affected and therefore to design appropriate cognitive measures. In their view, the lack of a standardized battery of cognitive tests has been a major stumbling block in this area of research. They propose that the way forward is to use practical measures, such as school attendance or performance in school examinations. There are drawbacks with this proposal, however.

Firstly, a delay in the behavioural and cognitive development of children or an impairment in cognitive function may be easier to identify than the biological causal mechanisms. If a developmental delay is found, it is important in its own right and may point the way to the underlying biological mechanism.

Secondly, batteries of tests do exist that are adequate for testing a variety of cognitive functions in different countries and cultures, but some aspects of child development are strongly influenced by both context and culture. Vocabulary tests are a case in point: terms and objects that are familiar to children in one society may be outside the experience of children in another. Skilled use of a language such as English, with a published literature, is associated with the use of literary terms; by contrast, skilled language use in a non-literate culture may be associated with other features of language, such as the effective use of metaphors, alliteration, or onomatopoeic expressions. Standard vocabulary tests may be insensitive to key language development factors in this instance.

Thirdly, the presence of endemic geohelminth infections implies an impoverished environment in which the sanitation services are generally inadequate. In these circumstances school attendance may be affected: children may be needed to assist in the household, to care for younger siblings, or to take part in agricultural tasks. Absenteeism and school drop-out rates are a reflection mainly of socioeconomic status (SES) and relatively insensitive to changes in child health.

Connolly (Connolly & Kvalsvig, 1993; Connolly, 1998;) and Watkins & Pollitt (1997) have recorded their views on the psychological functions most likely to be affected by parasites and the type of measures most likely to yield useful results. There are a number of candidate assessment measures:

- So-called "low-level" functions. In tests, subjects perform basic information-processing tasks such as searching and retrieving information from working memory. The speed with which the tasks are accomplished reflects the efficiency of these basic processes. These measures are regarded as the most promising for detecting biologically induced performance deficits.
- "High-level" functions, as measured by IQ and school performance, are more easily linked to learning and experience. In the long term these measures would be expected to reflect the constraints imposed on development by parasite infections, but they may not be sufficiently flexible to show a change in the short term following treatment.
- Instruments that track activity levels and tests of sustained performance over a long period (such as a vigilance task) could provide indication of problems with energy availability. Infected children may be able to respond adequately in a relatively brief test but may not be able to sustain attention or activity.
- Motivation and emotion. These two important aspects of children's behaviour have been neglected in studies so far.
- Rate of motor, language, social, and cognitive development. Almost all published studies focus on school-aged children, but children in endemic areas usually acquire parasitic infections at an earlier age. Almost nothing is known about how this affects their development or the possible consequences for educational performance later.

In the developmental progression outlined in Figure 1, it is obvious that the most rapid progress in acquiring motor and language skills takes place before average helminth prevalence rates are high in this community. However, averaging may obscure the real impact of first-time infections on immunologically immature children; a study in Jamaica of children suffering from dysentery caused by *T. trichiura* provided evidence of an effect on motor development (Callender et al., 1993). It is one of the central arguments of this report that scant attention has been paid to the developmental dimension. Parasitic infections are typically acquired at a time when the child is developing rapidly, and different species of parasitic infection may be acquired concurrently, successively, and continuously during what are colloquially termed "the formative years". The important point about these developmental milestones is that they do not occur in isolation from each other: cognitive constructs are built on the foundations of what has previously been experienced and understood. There is a danger of a cumulative effect on development in regions where children suffer from an unremitting sequence of parasitic infections.

Implications for public policies and programmes

Although understanding of the impact of parasitic infections on cognition remains imperfect, it is important to consider the implications for public policies and programmes. There are now several safe, effective, and inexpensive drugs available for treatment programmes (Albonico et al., 1999). In line with international thinking on the rights of children to special protection, and the current "globalization" debates which emphasize responsibilities that transcend national boundaries, it is now clear that the economics of the country concerned are no longer the main limiting factor in the control of parasitic infections. The question of whether to control parasites or not should be settled on other grounds.

From a psychologist's point of view, the *timing* of an intervention is an important consideration, and the literature on nutritional deficits indicates that early intervention is more advantageous (Pollitt, 1996). At this stage, however, it is not clear whether this principle applies to parasite treatment programmes. While it is known, for example, that the intensity of infections is lower before the age of 6 years, it is not known whether first-time infections may provoke a larger immune response and, consequently, be more developmentally disruptive than later "add-on" infections, or whether early, light infections have negligible effects. It appears that the treatments used are safe even at an early age, and so, until more information is available, it seems sensible to extend school-based treatment to 4–6-year-olds as well. Perhaps, to contain delivery costs, younger siblings could be invited to the schools and clinics on treatment days, or treatment could be given at nursery schools and other preschool facilities, where they exist.

Not only is the timing of the intervention an issue, but the *time frame* in which it operates is important if the value of deworming as a public health intervention is to be assessed. Part of the confusion over the findings of deworming studies stems from the lack of information on the potential benefits to children of being free, or nearly free, of worm infections for an extended period of time. Possibly, the time frame of most studies has been too short: it is entirely plausible that benefits may take months or years to become manifest and may then take an unexpected form.

The debate over the long-term benefits to children of preschool enrichment programmes has alerted us to the fact that small initial gains in adapting to primary school appear to "wash out" after a few years. Intervention groups then become indistinguishable from controls on academic performance, but there are other gains that only become apparent later. The weight of evidence from a variety of studies on graduates of the Perry Preschool Program, Head Start, and other programmes aimed at children from poor socioeconomic backgrounds in the United States shows that these programmes have enduring effects on educational performance and social competence (Zigler & Syfco, 1994). In the context of deworming programmes, a child who has been quiet and solitary because of lack of motivation or energy to participate socially may take time to develop social skills and to make friends once the limiting conditions have changed. An improvement in attention may similarly take time to translate into better class marks in school. Where schooling is dull, more energy may simply mean that a child pays more attention to games and less to schoolwork.

In order to make informed decisions about the allocation of resources to the most needy regions, coordinators of parasite control programmes need to know more about the relative importance of the different worm species and the intensities of infection that cause cognitive impairment. They also need this information to determine the frequency of treatment, particularly once a control programme is well established.

Parasite control programmes can be viewed as a way of protecting children against some of the ravages of poverty. Poverty is a confounding factor in any account of the effects of worm infections. Endemic worm infections are most prevalent and intense in the slums and shanty towns of developing countries (Crompton & Savioli, 1993); within poor communities, socioeconomic status is strongly associated with prevalence and intensity of parasite species. (Holland et al., 1988; Mangali et al., 1993) Malnutrition and micronutrient deficiencies are prevalent in the same areas, and are coupled with inadequate provision of health and educational services. Families living under these stressful conditions are more likely to be dysfunctional or violent. All of these environmental and social conditions have been severally and separately linked with increased risk of poor school performance or failure to continue with schooling (see Figure 2).

In efforts to achieve this protection for children, a primary health care approach emphasizes that parasite control cannot be accomplished with a single strategy such as routine treatment of schoolchildren. It is not just the business of the health department control teams or school health personnel but should be pervasive, reaching every level of health care and spreading across sectors. There are many varied opportunities for improving the control of parasite infections—nurses counselling mothers at clinics, health inspectors setting standards of hygiene, teachers promoting better hygiene practices in schools, and budgeting to ensure that adequate resources are available for the building and maintenance of latrines in schools. All such measures not only increase the effectiveness of parasite control programmes but also have many other beneficial consequences. Parasite control programmes that incorporate sanitation and health education components may

Figure 2. Contextual factors that predict poor educational progress in children from disadvantaged communities in KwaZulu-Natal

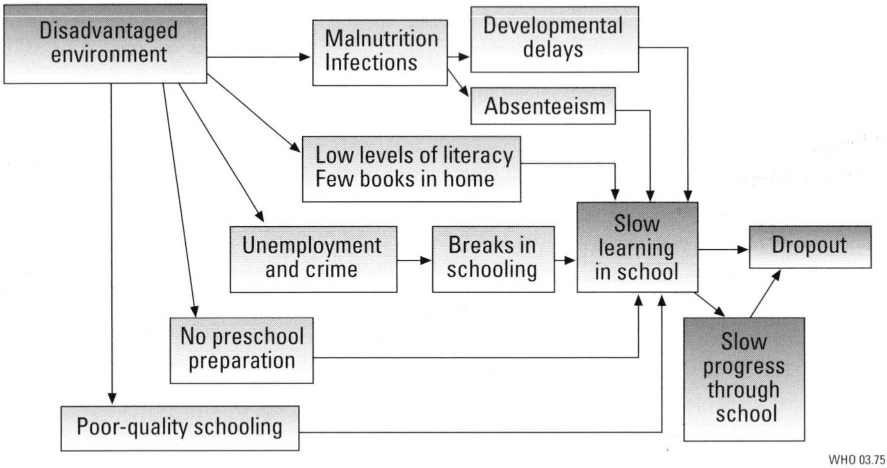

protect children against many other diseases, including opportunistic infections such as typhoid and cholera.

Control programmes that include lively and engaging health education provide schoolchildren with an understanding of the causes of ill-health and thus help them to safeguard their own health and the health of others. Child-to-child programmes use this opportunity to reach non-enrolled children and even parents with health messages. They take advantage of the fact that many schoolchildren are child-minders themselves and are well placed to practise and teach hygiene.

Current public health philosophy emphasizes the importance of holistic interventions such as WHO's Health-Promoting Schools Programme or the Integrated Maternal and Child Health initiatives. Deworming programmes are relatively simple to run and can provide entry points for more comprehensive health programmes linking health services to other governmental and nongovernmental agencies concerned with child development. They can be viewed as one component in a set of interventions designed to assist vulnerable children to achieve good health in its widest sense, which encompasses healthy development of social and cognitive competence (Dickson et al., 1999; Taylor et al., 1999).

References

Albonico M et al. (1999) A controlled evaluation of two school-based anthelminthic chemotherapy regimens on intensity of intestinal helminth infections. *International Journal of Epidemiology*, 28:591–596.

Callender J et al. (1993) Developmental levels and nutritional status of children with the *Trichuris* dysentery syndrome. *Transactions of the Royal Society of Tropical Medicine and Hygiene*, 87:528–529.

Carter R (1998) *Mapping the mind*. London, Orion.

Connolly KJ (1998) Mental and behavioural effects of parasitic infection. In: *Nutrition, health and child development: research advances and policy* recommendations. Washington, DC, Pan American Health Organization (PAHO Scientific Publication No. 566).

Connolly KJ, Kvalsvig JD (1993) Infection, nutrition and cognitive performance in children. *Parasitology*, 107(Suppl.):S187–S200.

Crompton DWT, Savioli L (1993) Intestinal parasitic infections and urbanization. *Bulletin of the World Health Organization*, 71:1–7.

Dickson R et al. (1999) Anthelminthic drugs for treating worms in children—effects on growth and cognitive performance. In: *Cochrane database of systematic reviews*, Issue 4.

Holland CV et al. (1988) Intestinal helminthiases in relation to the socioeconomic environment of Panamanian children. *Social Science and Medicine*, 26:209–213.

Mangali A et al. (1993) Intestinal parasitic infections in Campalagian district, south Sulawesi. *Southeast Asian Journal of Tropical Medicine and Public Health*, 24:313–320.

Nokes C, Bundy DAP (1994) Does helminth infection affect mental processing and educational achievement? *Parasitology Today*, 10:14–18.

Pollitt E (1996) Timing and vulnerability in research on malnutrition and cognition. *Nutrition Reviews*, 54(Suppl.):S49–S55.

Pollitt E et al. (1991) Effects of infections with *Trichuris trichiura*, *Ascaris lumbricoides* and hookworm on information processing among Kenyan school children. *FASEB Journal*, 5:A1081.

Sakti H et al. (1999) Evidence for an association between hookworm infection and cognitive function in Indonesian school children. *Tropical Medicine and International Health*, 4:322–334.

Sameroff AJ (1987) The social context of development. In: Eisenberg N, ed. *Contemporary topics in developmental psychology*. New York, Wiley: 273–291.

Stoltzfus RJ et al. (1998) Effects of the Zanzibar school-based deworming program on iron status of children. *American Journal of Clinical Nutrition*, 68:179–186.

Taren DL et al. (1987) Contributions of ascariasis to poor nutritional status in children from Chiriqui Province, Republic of Panama. *Parasitology*, 95:603–613.

Taylor M et al. (1999) Helminth control as an entry point for health promoting schools in KwaZulu-Natal. *South African Medical Journal*, 89:273–279.

Wachs TD (1996) Known and potential processes underlying developmental trajectories in childhood and adolescence. *Developmental Psychology*, 32:796–801.

Watkins WE, Pollitt, E (1997) "Stupidity or worms": do intestinal worms impair performance?" *Psychological Bulletin*, 121:171–191.

Werner EE (1989) Children of the Garden Island. *Scientific American*, 260:106–111.

Zigler E, Syfco SJ (1994) Head start: criticisms in a constructive context. *American Psychologist*, 49:127–132.

The effect of soil-transmitted helminth infection on the cognitive function of schoolchildren

S.A.N. Abidin,[1] P. Hadidjaja
Department of Parasitology, Faculty of Medicine
University of Indonesia, Jakarta, Indonesia

Introduction

The prevalences of soil-transmitted helminth infections in Indonesia are generally very high. It is estimated that the prevalence of *Ascaris lumbricoides* infection varies from 60% to 90%, especially in urban slum areas. Prevalence of *Trichuris trichiura* infection in the same areas is reported to be slightly lower, at 40–60%. By contrast, hookworm infection appears to be generally of very low prevalence—about 10% (Ismid et al., unpublished data); it is reported to be endemic only in plantation areas (e.g. tea and rubber plantations), where the prevalence among labourers is about 60%.

Any of these soil-transmitted helminth infections may result in morbidity, malnutrition, and iron-deficiency anaemia (Stephenson, 1987). Their possible contribution to impaired cognitive function and educational achievement has been demonstrated by the association between iron-deficiency anaemia and malnutrition in Indonesia (Soemantri, 1989). Indeed, a correlation between helminth infection and educational achievement has long been recognized (Nokes et al., 1991, 1992). This review paper discusses a number of reports on the correlation between helminth infections and cognitive function.

Soil-transmitted helminths and cognitive function

Research on the correlation between soil-transmitted helminth infections and cognitive function has focused on school-age children. Not only are these children the most vulnerable to helminth infections—they are also the population group most likely to experience the impact of infection on cognitive function. For hookworm infection, however, evidence for any direct correlation is very scarce.

Early in the 20th century, Stiles (1915a) undertook a study in the southern USA; he was probably the first to demonstrate the existence of a link between helminth infection and the educational achievement of schoolchildren. He found

[1] Professor S.A.N. Abidin, Department of Parasitology, Faculty of Medicine, University of Indonesia, Jalan Salemba Raya No. 6, Jakarta Pusat 10430, Indonesia
Tel: +62 21 310 2135, Fax: +62 21 391 4607

that children infected with hookworm, and to a lesser extent with *A. lumbricoides*, advanced through school more slowly than uninfected children, with an average deficit of 0.23 grades. The deficit was also correlated, however, with poor sanitary conditions. Because there was no recognizable impact of helminths on the memory span of infected children, Stiles (1915b) considered that the more frequent grade repetition among infected children was due to the impact of morbidity on school attendance and not the result of some direct effect on mental processing.

Waite & Neilson (1919) examined the relationship between intensity of hookworm infection and the intelligence quotient (IQ) of children in Queensland, Australia. They found that the degree of mental retardation increased in proportion to the intensity of infection and suggested that this was due to prolonged anaemia and toxaemia. However, no account was taken of potential confounding variables, such as socioeconomic status.

Nokes et al. (1991), investigating the effect of intensity of soil-transmitted helminth infection on cognitive function in children aged 9–11 years in three Jamaican schools, found a negative correlation between intensity and academic performance. Children with the least academic ability (as assessed by teachers from examination results) were more likely not only to be infected, but also to harbour above-average worm burdens. The level of school absenteeism was also related to infection. The proportion of the year during which these children were absent from school increased with increasing intensity of infection with *T. trichiura* (Nokes & Bundy, 1993): the more heavily infected individuals were absent almost twice as often as their uninfected peers.

Intervention study

The possibility of a causal association between helminth infections and educational achievement or cognitive function can be addressed through intervention studies. This approach has been used by several researchers who have measured the change or improvement in performance following drug treatment of the parasite infection.

In a double-blind placebo trial, Nokes et al. (1992) examined whether infection with *T. trichiura* was causally related to cognitive function in schoolchildren in Jamaica. For their study samples, they chose 104 children with moderate to heavy burdens of *T. trichiura*. The children were divided into three groups (treated, control, placebo) and assessed in a battery of eight tests administered both before and 63 ± 8 days after treatment with albendazole. Intra-observer reliabilities were assessed before the start of the study; a correlation coefficient of $r > 0.70$ was set as the minimum acceptable level of test repeatability for inclusion in the trial. These eight tests were chosen because, in factor analysis, they load on a "freedom from distractibility" factor. Only one person was in charge of their administration.

Three of the children's tests—Digit-Span Forwards/Backwards, Arithmetic, and Coding—were taken from the Wechsler Intelligence Scale. Other tests included the Matching Familiar Figures test, which involved problem-solving abilities, Listening Comprehension, involving memory and information processing, and Fluency, which involved long-term memory scanning and retrieval. To assess whether the randomization of treatment and placebo had successfully distributed possible confounding variables equally between the two groups, school attendance, IQ (using Raven Coloured Progressive Matrices), and social background characteristics were recorded at baseline. The treatment group (infected children) received a single 400-mg dose of albendazole following the initial cognitive testing. Control (uninfected) and placebo (infected) groups received the same placebo. The rates of improvement in the placebo and control groups were compared using Mann-Whitney *U*-test on the differences in slope. Multiple linear regression was also used to compare the treatment and control groups.

Multiple regression analysis of treatment and placebo groups showed that children who received anthelminthic treatment improved significantly more in three cognitive tests than those who received placebo, as indicated by the positive regression coefficient of the categorical variable treatment versus placebo. This significant effect of treatment was observed in the tests of Fluency ($P < 0.001$), Digit-Span Forwards ($P < 0.02$), and Digit-Span Backwards ($P < 0.01$). The effect of treatment was not significant in any other test. In addition, multiple regression analysis revealed that the treatment group improved significantly more than the control group in Fluency ($P < 0.003$), Digit-Span Forwards ($P < 0.05$), and Digit-Span Backwards ($P < 0.01$). It was concluded that the helminth infections affected cognition through their effects on the general well-being of the infected children. The fatigue and listlessness experienced by the children with moderate to heavy loads of *T. trichiura* might have been responsible for a suboptimal level of arousal (Nokes et al., 1991, 1992).

Hadidjaja et al. (1998) reported the effects of intervention on nutritional status and cognitive function of primary-school children infected with *A. lumbricoides*. The study was conducted in slum areas in which environmental sanitation, as well as personal hygiene, was poor. The researchers recruited approximately 1000 children who were targeted for parasitological, nutritional, and psychological examination before and after an intervention; the interval between these two examinations was 5 months. The objective of their research was to study the effect of *A. lumbricoides* infection on nutritional status and cognitive function. Because infection with *A. lumbricoides* alone is rare in Indonesia, mixed infections of *A. lumbricoides* and *T. trichiura* could not be avoided in the study population. However, children with high counts of *T. trichiura* eggs were excluded from the study because moderate to high intensity of infection with this worm could cause a reduction in cognitive function (Nokes et al., 1994).

The statistical procedures used were analysis of variance and covariance. All analysis was done using the Statistical Program for Social Sciences. In addition, Kolmogorov–Smirnov and chi-square tests were used to analyse nutritional and parasitological status. The interventions used were anthelminthic treatment with mebendazole and health education, and the groups studied were: (1) treatment with mebendazole, (2) health education alone, (3) health education in addition to treatment with mebendazole, (4) placebo, (5) uninfected control.

The results showed no significant difference in nutritional status before and after the interventions. Analysis of covariance indicated a significant effect of the different treatments only for the Coloured Progressive Matrices, Coding, and the number of eggs of both worms. In the Coloured Progressive Matrices, the greatest mean score was observed in the mebendazole-treated group; the mean was greater than those of the egg-negative and placebo groups. The test performance of the mebendazole-treated group, which had the lowest baseline level in this test, was clearly improved by treatment. This was also true for Coding.

Discussion

The impact of soil-transmitted helminth infections on cognitive function has been apparent from studies carried out by several research workers since the early 1900s. Only the findings of Stiles (1915a, 1915b) were not supported by tests to assess the intelligence of children. Stiles reported simply that children infected with hookworm advanced through school more slowly than uninfected children, with an average deficit of 0.23 grades. Waite & Neilson (1919) studied the relationship between the intensity of hookworm infection and the intelligence quotient of children in Queensland. They found that the degree of mental retardation increased in proportion to infection intensity and suggested that this was due to prolonged anaemia and toxaemia. Although the study did not consider potential confounding variables such as socioeconomic status, it revealed a clear correlation between the intensity of worm infection and cognitive function.

According to Nokes & Bundy (1994), an intervention study—in which change in performance following treatment of the parasite infection is measured—is the best method of assessing the causal association between helminth infection and cognitive function. Nokes et al. (1992) studied the correlation between *T. trichiura* infection and cognitive function of schoolchildren. They found that treatment of infected children aged 9–12 years with moderate to high *T. trichiura* burdens led to a significant improvement in auditory short-term memory and in scanning and retrieval of long-term memory 9 weeks after treatment. Hadidjaja et al. (1998) looked at the relationship between *A. lumbricoides* infection and cognitive function in primary-school children, and found that children treated with mebendazole showed significant improvement in the Coloured Progressive Matrices and Coding test. Children's learning ability, concentration, and hand–eye coordination also improved within 5 months of receiving the intervention.

Watkins, Cruz & Pollitt (1996) reported the effect of deworming on indicators of school performance in rural Guatemala. Of 246 children studied, 91% had ascariasis and 82% trichuriasis. These children were randomly designated to receive either albendazole or placebo at baseline and again 12 weeks later, and were followed for 6 months. Comparison of the treated and placebo groups showed no positive effect of deworming on reading, vocabulary, or attendance, but egg counts were lower in the treated group.

It is thus clear that there is a correlation between soil-transmitted helminth infections and cognitive function, but the mechanism by which worms affect cognitive function remains to be clarified. Some workers have postulated that undernutrition, with or without anaemia, is a contributory factor. Another possible explanation is the overall effect of the helminths on the general well-being of infected children. The fatigue and listlessness experienced by children suffering from worm infections may result in suboptimal arousal levels and impaired performance in tests. Further research is needed in this area.

References

Hadidjaja P et al. (1998) The effect of intervention methods on nutritional status and cognitive function of primary school children infected with *Ascaris lumbricoides*. *American Journal of Tropical Medicine and Hygiene*, 59:791–795.

Nokes C, Bundy DAP (1993) Compliance and absenteeism in school children: implications for helminth control. *Transactions of the Royal Society of Tropical Medicine and Hygiene*, 87:148–152.

Nokes C, Bundy DAP (1994) Does helminth infection affect mental processing and educational achievement? *Parasitology Today*, 10:14–18.

Nokes C et al. (1991) Geohelminth infection and academic assessment in Jamaican children. *Transactions of the Royal Society of Tropical Medicine and Hygiene*, 85:272–273.

Nokes C et al. (1992) Parasitic helminth infection and cognitive function in school children. *Proceedings of the Royal Society of London, Series B*, 247:77–81.

Soemantri AG (1989) Preliminary findings on iron supplementation and learning achievement of rural Indonesian children. *American Journal of Clinical Nutrition*, 50(Suppl.):698–702.

Stephenson LS (1987) Ascariasis. In: Stephenson LS, Holland C, eds. *The impact of helminth infections on human nutrition; schistosomes and soil-transmitted helminths.* London, Taylor & Francis:89–127.

Stiles CW (1915a) The school grades attained by 2166 white schoolchildren in the city of x, classified by age, sanitation, and intestinal parasites. *United States Public Health Reports*, 30:2060–2067.

Stiles CW (1915b) Memory tests of schoolchildren. *United States Public Health Reports*, 30:3738–3745.

Waite JH, Neilson L (1919) A study of the effects of hookworm infection upon the mental development of North Queensland school children. *Medical Journal of Australia*, 1:1–17.

Watkins WE, Cruz JR, Pollitt E (1996) The effects of deworming on indicators of school performance in Guatemala. *Transactions of the Royal Society of Tropical Medicine and Hygiene*, 90:156–161.

PART II

Strategies and major programmes for the control of disease due to helminth infections

Strategies and major programmes for the control of disease due to helminth infections

Successful, sustainable control of disease due to helminth infections must be founded on comprehensive planning. There are no short cuts: good planning is vital if effective implementation and monitoring are to be achieved. A great deal of experience has been accumulated concerning strategies for helminth control programmes; a selection of useful introductory literature is given below.

Suggested reading

Chitsulo L, Lengeler C, Jenkins J (1995) *The schistosomiasis manual.* Geneva, World Health Organization (document TDR/SER/MSR/95.2).

Montresor A et al. (2002) *Helminth control in school-age children: a guide for managers of control programmes.* Geneva, World Health Organization.

Otteson EA et al. (1997) Strategies and tools for the control/elimination of lymphatic filariasis. *Bulletin of the World Health Organization*, 75:491–503.

Strategies for the control of intestinal helminths and lymphatic filariasis: initiatives in the WHO Western Pacific Region

K. Palmer,[1] K. Ichimori
WHO Regional Office for the Western Pacific
Manila, Philippines

Introduction

Soil-transmitted helminth infections are a major health problem in many developing countries of the Western Pacific Region, yet for most of those countries there is very little useful information that describes the scope and intensity of the problem. It is unfortunate that these widespread diseases, for which relatively simple but effective control strategies exist, are often totally overlooked in favour of other more "fashionable" health problems.

The renewed interest in parasitic diseases, led by Japan's Hashimoto Initiative together with programmes such as "Health-Promoting Schools", has opened the way for new programmes that either have incorporated or will soon incorporate the control of soil-transmitted intestinal parasites. In the WHO Western Pacific Region, a combined campaign against lymphatic filariasis and intestinal parasites is planned in 22 Pacific island countries and territories.

Elimination of lymphatic filariasis in the Pacific

Lymphatic filariasis caused by *Wuchereria bancrofti* has had a major impact on the health of many populations of the small Pacific islands. At one time, survey results showed microfilaria infection rates of more than 40% in some countries. Recent surveys reveal much lower rates, but it is still common to find villages where there is active transmission and many older people with elephantiasis.

In May 1997, the World Health Assembly called for the global elimination of lymphatic filariasis as a public health problem by 2020; in March 1999, ministers and directors of health endorsed the development and implementation of a comprehensive strategy to eliminate lymphatic filariasis in all 22 Pacific island countries and territories. On the basis of that endorsement, the World Health Organization and the Secretariat of the Pacific Community sponsored a meeting

[1] Dr K. Palmer, World Health Organization, Regional Office for the Western Pacific, United Nations Avenue, P.O. Box 2932, Manila, Philippines
Tel: +63 2 528 9725, Fax: +63 2 521 1036, E-mail: palmerk@wpro.who.int

in June 1999 of public health practitioners and others working on filariasis in the Pacific. Out of that meeting came the formation of the Pacific Initiative for the Elimination of Lymphatic Filariasis (PacELF) coordinating body. Members of the coordinating body met first in December 1999 to endorse a Pacific-wide plan of action for filariasis elimination: the plan calls for elimination of lymphatic filariasis from the Pacific by 2010, i.e. 10 years before the global target date. It is an ambitious but attainable goal.

The PacELF strategy

PacELF exists to coordinate activities among 22 Pacific island countries and territories that themselves have few resources. It was formed in recognition of the fact that many people travel from island to island and country to country, so that the Pacific essentially resembles one large country with a total population of more than 7 million people. By working together and sharing resources, including information, the small countries and territories can help each other to implement a comprehensive regional strategy for elimination of filariasis and control of intestinal parasites.

The countries and territories have been divided into three groups based on available microfilaria rates as shown in Table 1.

- Group 1 consists of countries and territories that, according to available information, have no filariasis. Surveys are needed to confirm the status, and it is possible that some areas of active transmission may be found in the course of those surveys.
- Group 2 consists of small island countries and territories where there is known to be transmission and where there will be annual mass drug administration (MDA) of diethylcarbamazine (DEC) and albendazole to the entire population. Most of these countries started MDA in 2000 and 2001; the others are in the process of formulating plans (see Table 1).
- Group 3 are the larger countries where filariasis is a problem but where MDA activities will have to be carried out in stages and may not involve the whole country. Of the two countries in this group, Fiji already has an ongoing MDA programme and Papua New Guinea is still in the process of formulating a plan of action.

Some countries were delayed in starting the MDA activities planned for late 1999 and early 2000 because of delays in receiving approval for the use of the DEC + albendazole combination. Once approval was received, GlaxoSmithKline—the company that generously donated supplies of albendazole—rapidly shipped the drug to those countries that met the supply criteria.

The proposed regional timetable for PacELF activities is shown schematically in Figure 1.

The creation of PacELF represents a unique solution to public health inter-

Table 1. Grouping of Pacific island countries and territories, and planned dates for mass drug administration

Country or territory	Estimated population	Date of MDA
Group 1. Confirmation of filariasis status—total population 850 337		
Expected free		
Guam	147 300	
Northern Mariana Islands	65 100	
Marshall Islands	58 250	
Nauru	10 200	
Palau	17 250	
Pitcairn	47	
Tokelau	1500	
New Caledonia	196 840	
Solomon Islands	393 580	
Group 2. Pacific island elimination strategy—total population 930 730		
American Samoa	55 650	2000
Cook Islands	19 020	2000
Federated States of Micronesia	107 040	
French Polynesia	220 000	2002
Kiribati	77 800	2001
Niue	2 300	2000
Samoa	165 500	1999
Tonga	97 450	2001
Tuvalu	9 500	2001
Vanuatu	162 160	2000
Wallis and Futuna	14 400	2002
Group 3. Large-country elimination strategy—total population 4 999 180		
Fiji	772 660	2002
Papua New Guinea	4 226 520	

ventions among small island countries and is seen as the basis for implementation of many other programmes under the umbrella of "Healthy Islands".

PacELF is not a supplier, an administrator, a funding agency, or a technical authority. Its role is to serve as a coordinator and a centre for communications. It organizes meetings, serves as a clearinghouse for funding, and reviews and disseminates technical information. All 22 Pacific island countries and territories are members of PacELF. Within PacELF there is a central coordinating body (CB) made up of selected country representatives, one each from the Melanesian, Micronesian, and Polynesian countries, plus the two major partners—WHO and the Secretariat of the Pacific Community. The relationship between PacELF, the various donor partners, and the countries is shown in Figure 2.

Figure 1. Regional timetable for PacELF activities

		Pacific Region		
Step	**Year**	**Group 1 /CEF)**	**Group 2 (PIS)**	**Group 3 (LCS)**
Step 1	1999	Planning	Planning	Planning
	2000			
	2001		Intervention	
	2002	Evaluating		
	2003			Intervention
	2004		Evaluating	
	2005	**Country Elimination : Certificates**		
Step 2	2006	Planning		
	2007			
	2008	Evaluating		
	2009			Evaluating
	2010	**Regional Elimination : Declaration**		

WHO 03.57

Figure 2. PacELF organization structure

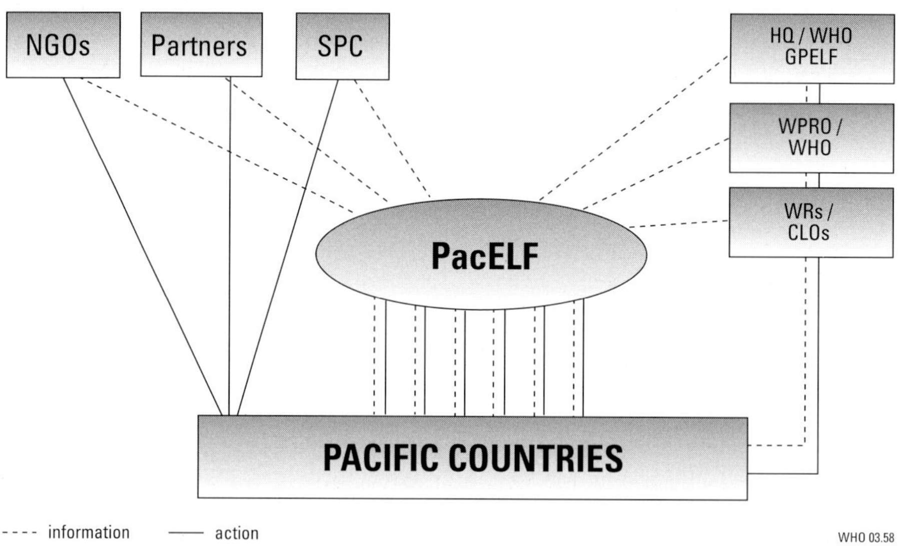

WHO 03.58

The collection of routine surveillance data and the formulation of elimination criteria suitable for use among small island populations are two major problems facing PacELF. Because of the nature of the disease, surveillance and assessment of the impact of elimination measures must be based on a system of active case-detection. This means regular surveys sampling the entire population at risk.

Established global criteria for filariasis elimination require a sample of 3000 children under 5 years of age to confirm the interruption of transmission (WHO, 1999). However, in small island populations there may not be 3000 children in that age group, and more work therefore needs to be done to define workable criteria for filariasis elimination in such populations; an entire population may have to be surveyed to determine whether transmission has been interrupted. For medium-sized populations of less than 50 000, it may be necessary to sample as much as half the population, while in larger countries the standard global sampling methods can be applied.

Samoa's MDA, in November 1999, achieved high coverage: more than 86% of the population took the single combined dose of DEC and albendazole. No adverse reactions to the drug combination were reported. American Samoa's MDA was completed in February 2000.

The annual distribution of albendazole will greatly reduce the intestinal worm burden (WHO, 1998). By the end of 2000 all countries in Group 2 plus Fiji planned to have completed one round of MDA, meaning that 1.8 million people would be treated for intestinal parasites. For those countries where filariasis had already been eliminated, an annual treatment of all schoolchildren was proposed as part of an active Health-Promoting Schools programme. Similar programmes involving schoolchildren were to be initiated in those areas of Papua New Guinea and Solomon Islands that would not be included in future MDA activities.

To assess the impact of albendazole on intestinal parasites, pre-MDA surveys have been carried out in Samoa and Vanuatu. Regular annual surveys of schoolchildren will be done to assess the impact of the annual MDA on such indicators as anaemia and stunting.

It is envisaged that other Healthy Islands projects, including those focusing on water supply and sanitation, diabetes control, the tobacco-free initiative, and related programmes, can be incorporated into the overall PacELF coordination framework.

References

WHO (1998) *Report of the WHO Informal Consultation on Monitoring of Drug Efficacy in the Control of Schistosomiasis and Intestinal Nematodes, Geneva, 8–10 July 1998.* Geneva, World Health Organization (document WHO/CDS/CPC/99.1).

WHO (1999) *Report of a WHO Informal Consultation on Epidemic Approaches to Lymphatic Filariasis Elimination: Initial Assessment, Monitoring, and Certification, Atlanta, Georgia, 2–4 September 1998.* Geneva, World Health Organization (document WHO/FIL/9.195).

The Hashimoto Initiative and the control of soil-transmitted helminth infections: the Japanese challenge

Tsutomu Takeuchi,[1]
Department of Tropical Medicine and Parasitology, School of Medicine, Keio University, Tokyo, Japan

Takeshi Kasai
Infection Control Division
Japanese Ministry of Health and Welfare, Tokyo, Japan

Takaaki Hara
Japan Association of Parasite Control, Tokyo, Japan

Introduction

Soil-transmitted helminth infections such as ascariasis, hookworm disease, and trichuriasis occur widely throughout the developing countries. Children and pregnant women are the main sufferers from these parasitic infections. Nevertheless, countermeasures have been limited compared with, say, efforts to control malaria. The estimated number of infected individuals worldwide is still in excess of 2 billion, and in some regions of endemic countries the infection rate exceeds 90%. Clearly, soil-transmitted helminth infections remain an enormous public health problem in the 21st century. In addition, these infections constitute a significant barrier to socioeconomic development in endemic countries.

These considerations underline the urgent need for implementation of control measures. In this regard, the success of helminth control efforts in several countries is particularly noteworthy. Safe and effective drugs are now available, and much valuable experience of promoting active community participation in helminth control has been gained. Suppression of soil-transmitted helminth infections has thus become a feasible objective—and one that has been achieved in Japan.

In June 1997, at the summit meeting of the heads of government of the G8 group of countries, held in Denver, USA, Mr Ryutaro Hashimoto, then Prime

[1] Professor Tsutomu Takeuchi, Department of Tropical Medicine and Parasitology, School of Medicine, Keio University, 35 Shinanomachi Shinjuku, Tokyo, Japan
Tel: +81 3 3353 1211 ext 62666, Fax: +81 3 3353 5958, E-mail: takeuchi@sc.itc.keio.ac.jp

Minister of Japan, pointed out the serious burden of parasitic disease in developing countries and stressed the need for international cooperation to reduce it. With the agreement of the other heads of government, Mr Hashimoto undertook to gather information on the current global status of parasitic diseases and, in the light of Japanese experience on parasite control, to compile a report for further discussion at the next summit meeting.

The Japanese Ministry of Health and Welfare subsequently convened a working group. In May 1998, after a series of studies and discussions, the working group published a report entitled *The Global Parasite Control Strategy for the 21st Century*. At the next summit meeting, in Birmingham, England, Mr Hashimoto made the concrete proposal—based on that report—that the G8 group of countries should take a leading role in parasite control efforts as a major contribution to world health.

During discussions at the summit meeting, Mr Hashimoto's proposal was endorsed by the other heads of government; a resolution reflecting this was incorporated into the official communiqué, with the statement that control of parasitic diseases, including malaria, should be implemented. The proposal has been subsequently referred to as the Hashimoto Initiative. The Japanese Government, the Japanese Society of Parasitology (JSP), and the Japan Association of Parasite Control (JAPC) organized a coordination committee for the Initiative, also known as the "global parasite control initiative", with the aim of promoting control strategies based on years of experience of parasite control in Japan.

Strategies and current activities of the Hashimoto Initiative

The Hashimoto Initiative identified four basic strategies as the foundation for its activities. These were stated as follows:

Strategy 1 Effective international cooperation for the efficient implementation of parasite control.

Strategy 2 Active pursuit of research that provides a scientific basis for parasite control.

Strategy 3 Active implementation of effective parasite control projects.

Strategy 4 Strengthening of the G8 countries' capabilities to deal with parasitic diseases—meaning that each of the G8 countries is requested not only to promote education and research on parasitic diseases but also to strengthen its ability to make a contribution to parasite control.

Japan has organized its own activity on the basis of these four strategies. The principal participants have been the Ministry of Foreign Affairs with the Japan International Cooperation Agency (JICA), and the Ministry of Health and

Welfare and Ministry of Education. In addition, JSP and JAPC have been involved in the planning and implementation of the Initiative.

Japan's current activities within the Hashimoto Initiative can be summarized as follows. First, the Ministry of Education has been promoting education in parasitology and related fields, mainly in medical schools. The Ministry of Health and Welfare has increased the grant for parasitic disease research. For example, this Ministry has supported various investigations into emerging and re-emerging infectious diseases, including malaria, schistosomiasis, amoebiasis, echinococcosis, and cryptosporidiosis; it has also provided support for a research group considering possible training modules for use at research and training centres (see below). In addition—and in collaboration with JSP and JAPC—it has been holding international workshops to expand coordination with international activities such as those undertaken by the World Bank and WHO. Financial support to multilateral organizations has also been expanded within the scope of the Initiative. Three training courses have been organized by JICA for parasite control personnel at managerial level; trainees have been invited from Africa and Asia.

The most significant activity of the Hashimoto Initiative is the establishment of research and training centres for parasitic disease control in Asia and Africa, which are run by JICA. These centres focus on the basic concept of the Initiative, making the best use of Japanese experience in parasite control. As is well known, Japan has been successful in eradicating malaria, schistosomiasis, and filariasis and has effectively suppressed soil-transmitted helminth infections.

Effective coordination between academic bodies, central/local governments, and nongovernmental organizations characterized Japanese efforts to control soil-transmitted helminthiasis. A school-based approach was adopted, combining health education and deworming, and proved to be highly effective for community mobilization, leading eventually to better community health. If research within the Hashimoto Initiative can successfully adapt this approach for use in developing countries, it will have made a significant contribution to improving community health. In view of the extensive theoretical and practical experience gained by Japan in the control of soil-transmitted helminth infections, there is a strong argument that these infections should be among the primary targets of human resource development at the research and training centres, which is another key activity of the Initiative.

Human resource development at the research and training centres

In 1998, Mr Hashimoto announced the establishment of three research and training centres—two in Africa and one in Asia—with the aim of contributing to parasitic disease control through human resource development. After preliminary studies, it was decided to base these centres at the Kenya Medical Research Institute (KEMRI) in Nairobi, the Noguchi Medical Institute of Ghana, and Mahidol

University in Thailand. Planning of the two African centres by JICA and discussions with each country are still continuing. The agreement for the establishment of the Asian centre—now called the Asian Centre of International Parasite Control (ACIPAC)—was officially signed between Mahidol University, the Government of Thailand, and JICA in March 2000.

During planning of the ACIPAC project, a preliminary study team was dispatched to Thailand in October 1999 to consider the purposes and characteristics of the Centre. After extensive discussion, the study team and their Thai colleagues agreed that the primary focus of the Centre should be capacity-building among managers and senior managers of parasite control programmes and among mid-level health personnel. The capacity-building programme should combine practical activities, similar to those conducted during technical co-operation schemes, the Third Country Training Programme, and the region-wide technical cooperation scheme. The training, technical support, operational research, and information networking functions of ACIPAC would be achieved with the assistance of JICA (see Figure 1).

The administration of the ACIPAC project was agreed by the Japanese and Thai collaborators; during the first year, "on-the-job" training was conducted as part of the JICA country project. This was followed during 2001 by the Third

Figure 1. Concept and strategy of the research and training centres in Ghana, Kenya, and Thailand

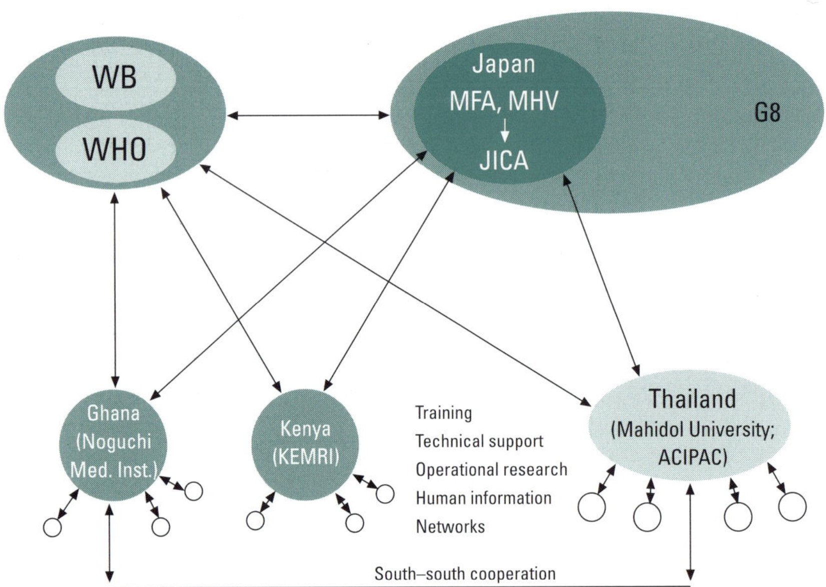

WB: World Bank
MFA: Ministry of Foreign Affairs
MHW: Ministry of Health and Welfare
JICA: Japan International Cooperation Agency
KEMRI: Kenya Medical Research Institute
ACIPAC: Asian Centre of International Parasite Control

WHO 03.76

Country Training Programme in which trainees from four surrounding countries (Cambodia, the Lao People's Democratic Republic, Myanmar, and Viet Nam) were trained by Thai scientists in collaboration with Japanese experts. Soil-transmitted helminth infections and malaria were the primary targets of training. Operational research methods were applied to improve the quality of the educational methodologies and of the parasite control programmes. The Centre also acts as an information network, facilitating contact between former students and international organizations, and, more importantly, between former students from different countries ("south–south cooperation"), with exchanges of information, technology, and human resources. For this purpose, cooperation with SEAMEO-TROPMED and other organizations such as ACT-MALARIA and JAPC's training course are being sought.

Although the ACIPAC project is essentially a bilateral programme between Japan and Thailand, the possibility of collaboration with international organizations and other bilateral donors is now being considered. Technical cooperation across the region should maximize the benefits of training. Training schemes could be established in other countries with support from multilateral and other bilateral donors.

Future activities of ACIPAC may be expected to include identification of basic research needs, and use of the Centre for training parasitologists and related personnel from Japan and other developed countries. More importantly, the Centre may be able to act as the "launch pad" for parasite control programmes, including a large-scale soil-transmitted helminth control programme, which is one of the mid- to long-term goals of the Hashimoto Initiative.

Conclusion

The Hashimoto Initiative was launched in response to a proposal made by the Prime Minister of Japan; consequently, its approval guaranteed it the highest level of political commitment. Active support is being sought from international communities. It is hoped that many parties and organizations will participate in the Initiative. Although their interests, purposes, and obligations may vary, a common understanding of the ultimate objective—reduction in the burden of parasitic diseases—will allow them to cooperate in working towards that objective and, as the Japanese saying goes, "to dream different dreams in the same bed".

A deworming programme in Nepal supported by the World Food Programme

Gian Pietro Bordignon,[1] Deepak Raj Shakya
School Feeding Programme Team
World Food Programme, Kathmandu, Nepal

Background

In June 1996, a survey was conducted jointly by WHO, the World Food Programme (WFP), and Nepal's Ministry of Health to evaluate the prevalence and intensity of infections due to soil-transmitted nematodes among schoolchildren in Nepal. Its aim was to evaluate the need for control measures for intestinal worms within the ongoing School Feeding Programme supported by WFP and implemented by the Ministry of Education.

The survey covered 711 primary-school children in Parsa and Dailekh districts and found an overall prevalence of worms of 90% in children. Stools were analysed with the Kato–Katz technique (see Table 1). This very high prevalence of intestinal worm infections, especially hookworm infection, is probably associated with severe iron-deficiency anaemia and with poor environmental hygiene. The survey also revealed the widespread prevalence of hookworms in rural areas and of whipworms in urban areas. Of the schoolchildren examined, 20% were either moderately or heavily infected with hookworms; boys were more heavily infected than girls in all the districts surveyed. Hookworm infection is a known cause of severe nutritional deficits, including iron-deficiency anaemia, and intestinal parasitic infections and anaemia rank as the fourth and fifth priority health problems in both rural and urban areas. The data collected during the survey confirmed other published findings: Nepal is one of the countries where soil-transmitted nematode infections are most important.

Recommendations made at the end of the survey stressed the urgent need for the following school-based control activities:

- regular drug distribution to school-age children;
- health education, concentrating particularly on soil-transmitted nematodes; and
- as a long-term objective, the improvement of sanitation in schools and villages.

---

[1] DR G.P. Bordignon, World Food Programme—West Africa Regional Bureau, P.O. Box 6288, Dakar, Senegal (present address)
Tel: +221 849 6500, fax: +221 842 3562, E-mail: gianpetro.bordignon@wfp.org

Table 1. Baseline parasitological survey in schoolchildren (Parsa and Dailekh districts, Nepal, June 1996)

Sample	
Total number of schoolchildren investigated:	711
Mean age (years)	9.265
Sex distribution (male : female)	54 : 46

Soil-transmitted helminth	Prevalence (%)	Mean intensity (eggs/gram)	Intensity class		
			Light (%)	Moderate (%)	Heavy (%)
Ascaris lumbricoides	21.9	815	17.7	3.9	0.3
Trichuris trichiura	19.2	38	18.8	0.4	—
Hookworms	64.7	1223	44.7	11.0	9.0
Any	74.2				

The survey emphasized that the establishment of control activities would strengthen the benefits of WFP's School Feeding Programme in terms of improving children's health, nutritional status, and school performance through the reduction of intestinal parasitic infection. Through the School Feeding Programme, WFP provides a hot midday snack—haluwa—which is made from sugar mixed with blended flour (consisting of wheat, maize, and soya bean), fortified with vitamins and minerals, and then mixed with vegetable ghee. The daily ration received by each child contains 85 g of blended flour, 15 g of sugar, and 10 g of vegetable ghee; it provides 17 g of protein, 15.1 g of fat, vitamin and micronutrient supplements, and a total of 463 $kcal_{th}$.[1] The area covered by the School Feeding Programme is shown on the map in Figure 1.

Objectives

In accordance with the findings of the survey, the WFP project "Food Assistance to Primary Schools" is providing two doses of deworming tablets per year to schoolchildren in the 12 districts covered by the project (out of the 75 districts that make up Nepal). Some 250 000 children enrolled in approximately 2500 public primary schools, mostly in rural and hill areas, are targeted. The objective is to strengthen the educational and nutritional benefits of the project by reducing the intensity and prevalence of intestinal parasitic infections in children aged 6–12 years. Deworming is intended to reduce the prevalence of hookworm, *Ascaris*, and *Trichuris* infections by 80% and the intensity of infection by 90%.

The planned district coverage and targeted beneficiaries are as follows:

- 10 districts during the 1998/1999 academic year—a total of 200 000 beneficiaries;

[1] 1 $kcal_{th}$ = 4.184 kJ.

Figure 1. Map of Nepal showing WFP-assisted project districts—School Feeding Programme

- 12 districts from 1999/2000 to mid-2002 (the end of the project phase)—a total of 250 000 beneficiaries.

According to Ministry of Education statistics, there were 23 284 primary schools with 3.46 million pupils in Nepal in 1997. The deworming programme was therefore expected to cover about 7% of the total primary school population in 1999/2000.

Funding for the programme

The deworming programme was funded with the support of the Canadian Women's Health and Micronutrient Facility; total budget allocation for the three-year period 1998–2000 was US$ 129 000.

Review of progress in 1998 and 1999

Procurement of drugs

As recommended by WHO, albendazole tablets (400 mg) were used for the deworming programme. Quotations for supplying the tablets were received from various sources, with the lowest (US$ 0.56 per tablet in 1998 and US$ 0.35 in 1999) coming from a manufacturer in Nepal. Samples of the tablets from this source were tested in Nepal and by WHO in Geneva in 1998; the results confirmed the good quality of the drugs.

Training of trainers and preparation of training materials

WHO provided technical assistance in preparing training materials and conducting trainers' training workshops before drugs were administered to schoolchildren. In May and June 1998, Dr A. Montresor from WHO in Geneva conducted training workshops for trainers in four project districts (Doti, Parsa, Rupandehi, and Surkhet), assisted by a Ministry of Health official and WFP staff from Kathmandu. Two District Health Officials and two staff from the Primary School Nutritious Foods Project (PSNFP)/Ministry of Education from each of the 10 project districts involved in deworming (Achham, Dadeldhura, Dailekh, Dhanusa, Doti, Parsa, Rupandehi, Rukum, Salyan, and Surkhet) participated in the training. These trainers, in turn, provided training to schoolteachers and parents in 1998. The District Health officials also provided training to schoolteachers and parents before tablets were distributed in the two new project districts in 1999.

Printing and distribution of posters

Posters illustrating (a) how worms develop in human body, (b) how worms can kill people, and (c) ways of preventing worm infection, plus flash cards depicting sanitation, were printed for training and distribution to schools. The designs for these posters and flash cards were received from UNICEF, WHO, the Ministry of Health, and the Family Planning Association of Nepal. In all, 3200 sets of posters (each of three pieces) and 3000 sets of flash cards were printed. Sets of these posters, flash cards, and other training materials were distributed to all the targeted primary schools in the 12 project districts.

Distribution of drugs

After receiving training, PSNFP, the agency collaborating in implementation of the Project, supplied the tablets to the primary schools through their normal distribution centres together with the regular food allocation. The trained schoolteachers then administered the first dose of albendazole tablets to schoolchildren in eight project districts (as opposed to the planned 10) in mid-1998; a second distribution round, covering approximately 200 000 primary-school children, was organized in 1999. Distribution of the second dose of tablets was delayed by administrative and logistic problems and by a delay in extension of the project to the two new districts in 1999. Distribution of two doses of tablets to 250 000 students in all 12 project districts was then conducted in 2000 and 2001 (one dose in the first quarter and the second in the last quarter).

Monitoring and evaluation

WHO provided technical assistance for monitoring the progress of the project. Forms were developed for completion by schools, PSNFP, and district health officials for reporting on training and drug distribution, and by parasitological teams for school survey/laboratory test reporting. Assistance in technical matters and logistics was solicited from the Nepalese Ministries of Health and Education.

Budget and expenditure

By the end of 1999, US$ 50 205 had been spent or committed for the two years 1998 and 1999, broken down as follows:

	US$
In 1998	
Procurement of 400 000 albendazole tablets	21 212
Printing of training materials and posters	4 199
Training and miscellaneous expenses	5 294
In 1999	
Procurement of 500 000 albendazole tablets	17 500
Training expenses	2 000
Total expenditure	*50 205*

Of US$ 129 000 received for this programme, a balance of US$ 78 795 remained at the end of 1999.

Results of the programme

Informal feedback received from several schools that distributed the deworming tablets revealed a significant improvement in pupils' health and nutritional status. Attendance in classes also improved. An impact assessment study planned for the end of 2000 would reveal the outcome of this programme intervention in more detail.

Expected outcome

A survey conducted in the areas covered by deworming in 2000 revealed a dramatic reduction in infection. The prevalence of worm infection was reduced from 74% to 51% and of high-intensity infection from 9.3% to 1.9%. Heavily infected individuals suffer most of the consequences of the infections and are also the major source of infection for the community; reducing the number of children in this group is therefore an important indicator of success. Significant nutritional improvements were achieved as a result of the control of soil-transmitted helminths: the percentage of anaemic children was only 11% in the sample when the expected prevalence was more than 50%.

Expansion of the deworming programme

On the basis of the positive impact of the WFP-assisted deworming programme, the Nepalese Ministry of Health (MOH), with the support of UNICEF, launched a new deworming intervention targeted at children aged 2–5 years in 14 districts during 1999–2000. Most of these districts were in the eastern region of the country, but three were districts already covered by WFP's School Feeding Programme in the central and western regions. The MOH deworming programme became an additional component of the continuing vitamin A supplementation programme already implemented in the districts concerned. In 2002, 48 of the 75 districts in Nepal integrated deworming with vitamin A distribution, covering more than 1.3 million children. The current level of interest and operational collaboration achieved between the MOH and the Ministry of Education, already tested during the WFP-supported deworming programme, augurs well for the continuation of the deworming programme and its future expansion in the primary education sector.

Partnership for Child Development: an international programme to improve the health of school-age children by school-based health services including deworming

Satoto,[1] Sri Hendratno, Herbanto Subagyo, Zen Rafilluddin, Michael Mogensen, Andrew Hall, Don Bundy
Partnership for Child Development
Department of Infectious Disease Epidemiology
Imperial College School of Medicine
London, England

Today, more than 90% of children in developing countries survive to school age, yet school-age children still suffer from a large disease burden. In many African countries, for example, half the children are stunted, suffer from anaemia and deficiencies of essential micronutrients, and are infected with parasitic worms. In addition to their detrimental physical effects, all of these conditions affect cognition and learning and may prevent children from taking full advantage of their limited opportunities for formal education (Bundy & Guyatt, 1996).

In its *World Development Report 1993: Investing in health*, the World Bank identified school health and nutrition programmes as one of five priorities for public health initiatives. Theoretical analysis had indicated that the health of schoolchildren could be improved at very low cost by a package of simple services, such as anthelminthics, micronutrients, and health education, delivered to children through the existing school system. However, there was little prior experience of either large-scale programmes or operational research on school health. In response to this need, the Partnership for Child Development was established in 1992 and set out to evaluate this package of services in four partner countries (Ghana, Indonesia, United Republic of Tanzania, and Viet Nam) on an operationally realistic scale.

The challenge in designing health programmes for school-age children is to ensure that the programmes can be sustained. Many of the most prevalent health problems of the school-age child can be alleviated at remarkably low cost. The available interventions are both effective and cost-effective, and the school setting

[1] Dr Satoto, University of Diponegoro, Semarang, Central Java, Indonesia
Tel/fax: +62 24 746 0038

offers an established and comprehensive system for health service delivery: there are more teachers than nurses, more schools than clinics. The education sector provides a substantial infrastructure for reaching schoolchildren in developing countries, and careful planning of simple health interventions will ensure that teachers are not overburdened (Bundy & Guyatt, 1996).

In each of the four partner countries, a counterpart organization was identified to coordinate programmes and work with the ministries of health and education to implement simple and cost-effective programmes. The four countries had different priorities for health interventions, but each had a core package of anthelminthic treatment and participatory health education. The education and health sectors of each country worked together to deliver this package, while local technical institutions conducted annual surveys to monitor the impact and cost of the intervention.

One of the four core programmes of the Partnership, the Indonesia Partnership for Child Development, *Mitra*, was set up in 1995 in the Central Java Province, principally in Karanganyar and Wonogiri districts. The programme was designed to deliver albendazole and health education to schoolchildren through *Usaha Kesehatan Sekolah*, an existing school-based health programme. Overall coordination of the programme was undertaken by the policy office of the Ministry of Health, and monitoring and evaluation were carried out by a team from the Diponegoro University Research Institute, coordinated by Dr Satoto.

A baseline study of the effects of worm infection on cognitive function and motor skills of schoolchildren in Karanganyar District, using battery tests, revealed a significant difference in school performance between infected and uninfected pupils. The baseline study was followed by mass treatment of 55 000 primary school-age children with albendazole. Drug treatment was reinforced by health education. The other half of Karanganyar District served as a comparison area for the first year of monitoring. After 1 year, children were re-examined to assess the impact of both deworming treatment and health education. In the second year of the programme, a baseline survey was carried out in schools in one half of Wonogiri District, followed by mass treatment in the third year.

A stepped design was chosen in which the programme is implemented in successively larger areas over a period of 4 years. This design met the needs both of implementation, which is scaled up in each successive year, and of monitoring and evaluation, in that children in new areas, who have not received treatment, are compared with the children in the first treatment area who have received successive years of treatment.

The annual costs of albendazole treatment was subsidized in a stepwise fashion. *Mitra* paid the full cost of treatment in the first year and half the cost in the second year; after that the full cost had to be met locally. This enabled the participating schools to sustain the programme over time. A school health insurance scheme was proposed to provide support for treatments as the subsidy was withdrawn and to promote local sustainability of the programme and other health services.

The health education component of the programme included an initial survey of knowledge, attitudes, and practices among pupils in primary schools in the district. The results were used as a basis for the planning and design of relevant materials to be distributed to both pupils and teachers. Among these materials were several booklets on worm infestation and its control that used a comic strip format to inform children of how infestation occurred and what could be done to prevent it, as well as the importance of environmental health for staying healthy.

Through its Indonesia and other country programmes, the Partnership has identified inexpensive delivery mechanisms and developed innovative ways to implement school-based health services. Based on the practical experience gained in the programmes, it has been able to show that there is a clear need for school-based health programmes, that the tools to implement such programmes are cost-effective, and that this type of programme is therefore a viable option in low-income communities.

The Indonesia programme used a simple questionnaire administered to pupils by their teachers. This approach used self-reporting as a reliable indicator of infection with *Schistosoma haematobium*, which is a highly cost-effective way of identifying both schools that warrant mass treatment for schistosomiasis and infected individuals within schools that do not warrant mass treatment (Partnership for Child Development, 1999).

Another example of an inexpensive and innovative delivery mechanism (although not used in the Indonesia programme) is provided by height (or tablet) poles. Most anthelminthics are safe, effective, and simple to give. WHO has recommended mass treatment of all schoolchildren if the prevalence of worm infections is greater than 50%. The drug most widely used to treat schistosomiasis is praziquantel, given as a single dose of 40 mg/kg body weight. However, the need to adjust the dose of praziquantel for body weight adds a significant complication to school-based treatment programmes: weighing scales are relatively expensive and easily damaged. If teachers are to administer praziquantel to their pupils as a part of school-based programmes, a simple and inexpensive alternative method of determining the correct dose is needed (Hall et al., 1999). Research has shown that height can be used as a proxy for weight, allowing the use of a height pole in place of scales for determining the correct dosage of praziquantel. Each child to be treated can simply be stood against a pole marked with the number of tablets or sections of tablet of praziquantel appropriate for his or her height (Hall et al., 1999).

The four country programmes have also shown the potential impact of school health programmes on the school-age child. Evaluation of large-scale demonstration programmes in both Ghana and the United Republic of Tanzania has shown that school-based health services can have an impact on a broad range of health and education outcomes. For example, in the United Republic of Tanzania, a significant increase in height (15 mm over 16 months) and haemoglobin levels (4.8 g/l) was observed in treated children (Beasley et al., 1999).

The Partnership is now seen as a leading agency providing technical assistance, quality assurance, and advice on programme design to governments, agencies, and others who wish to implement effective school health programmes. Its activities have included both work and assistance or advice to child health programmes in: Argentina, Australia, Bangladesh, Botswana, Cameroon, Chile, China, Colombia, Dominica, Egypt, Ethiopia, India (Uttar Pradesh), Jamaica, Kenya, Malawi, Malaysia, Mexico, Montserrat, Panama, Peru, South Africa, Sri Lanka, Saint Lucia, Thailand, Uganda, and Zambia. The Partnership collaborates on school health programmes with nongovernmental organizations in 9 countries, with bilateral agencies in 12 countries, with multilateral agencies in 37 countries, and with industry in 10 countries.

The Partnership for Child Development is promoting a school health dissemination and communications initiative with partner agencies and school health practitioners. The initiative includes the publication of reports and papers in scientific journals, the design of and support for school health workshops and training courses, and a school health web site and mailing list. The Partnership is also an active member of the FRESH Start Partnership (Focusing Resources on Effective School Health), launched at the World Education Forum in Senegal in April 2000. The FRESH Start partnership has outlined a core group of cost-effective activities that could form the basis for intensified and joint action to make schools healthy for children and so contribute to the development of child-friendly schools.

Children in low-income countries still face health problems that compromise their physical development, their attendance at school, their ability to learn, and—in many cases—their only chance for a formal education. Opportunities to both reduce the burden of childhood disease and equip children with the health education and life skills necessary to remain healthy and safe cannot be missed.

The Scientific Coordinating Centre of the Partnership for Child Development is located at the Imperial College School of Medicine in London and acts as a resource base for workers in the field of school health throughout the world. Information about Partnership activities and publications is available through the web site (http://www.child-development.org). The Partnership has also set up a school health mailing list and a web site (http://www.schoolsandhealth.org), in partnership with, among others, the World Bank, WHO, the United States Agency for International Development (USAID) Bureau for Africa Office of Sustainable Development Education Team, and the USAID Bureau for Latin America and the Caribbean Education Team.

References

Beasley NMR et al. (1999) The impact of population level deworming on the haemoglobin levels of schoolchildren in Tanga, Tanzania. *Tropical Medicine and International Health*, 4:744–750.

Bundy DAP, Guyatt HL (1996) Schools for health: focus on health, education and the school-age child. *Parasitology Today*, 12:1–15.

Hall A et al. (1999) Alternatives to bodyweight for estimating the dose of praziquantel needed to treat schistosomiasis. *Transactions of the Royal Society of Tropical Medicine and Hygiene*, 93:653–658.

Partnership for Child Development (1999) Self-diagnosis as a possible basis for treating urinary schistosomiasis: a study of schoolchildren in a rural area of the United Republic of Tanzania. *Bulletin of the World Health Organization*, 77:477–483.

Focusing resources on effective school health: a FRESH start to improving the quality and equity of education

D.A.P. Bundy[1]
World Bank, Washington, DC, USA

Background

"Education for all" means ensuring that all children have access to basic education of good quality. This implies creating in schools and in basic education programmes an environment in which children are both able and enabled to learn. Such an environment must be friendly and welcoming to children, healthy for children, effective with children, and protective of children. The development of child-friendly learning environments is an essential part of the overall efforts by countries around the world to increase access to, and improve the quality of, their schools.

Poor health and malnutrition are important underlying factors in low school enrolment, absenteeism, poor classroom performance, and early school drop-out, as reflected in the *World Declaration on Education for All*.[2] Programmes to achieve good health, hygiene, and nutrition at school age are therefore essential to the promotion of basic education for all children.

Good health and nutrition are not only essential inputs but also important outcomes of good-quality basic education. First, children must be healthy and well nourished in order to participate fully in education and gain its maximum benefits. Early childhood care programmes and primary schools that improve children's health and nutrition can enhance learning and educational outcomes. Second, education of good quality can lead to better health and nutrition outcomes for children—especially girls—and thus for the next generation of children as well. Moreover, a healthy, safe, and secure school environment can help to protect children from health hazards, abuse, and exclusion.

Positive experiences by the World Health Organization (WHO), the United Nations Children's Fund (UNICEF), the United Nations Educational, Scientific and Cultural Organization (UNESCO), and the World Bank suggested that a core group of cost-effective activities could form the basis for intensified and joint

---

[1] Dr D.A.P. Bundy, School Health and Nutrition, Human Development Network, The World Bank, 1818 H Street NW, Washington, DC 20433, USA.
Tel: +1 202 473 3636, E-mail: dbundy@worldbank.org

[2] Available on the Internet at www.unesco.org/education/efa/ed_for_all/background/jomtien_declaration.shtml.

action to make schools healthy for children and so contribute to the development of child-friendly schools. These agencies developed a partnership for Focusing Resources on Effective School Health, launching this FRESH Start approach at the World Education Forum in Senegal, in April 2000.

Ensuring that children are healthy and able to learn is an essential component of an effective education system. This is especially relevant to efforts to achieve education for all in the most deprived areas. Increased enrolment and reduced absenteeism and drop-out bring more of the poorest and most disadvantaged children, many of whom are girls, to school. It is these children, who are often the least healthy and most malnourished, who have the most to gain educationally from improved health. Effective school health programmes that are developed as part of community partnerships provide one of the most cost-effective ways to reach adolescents and the broader community and are a sustainable means of promoting healthy practices.

Improving the health and learning of schoolchildren through school-based health and nutrition programmes is not a new concept. Many countries have school health programmes and many agencies have decades of experience. These common experiences suggest an opportunity for concerted action by a partnership of agencies to broaden the scope of school health programmes and make them more effective. Effective school health programmes will contribute to the development of child-friendly schools and thus to the promotion of education for all.

This interagency initiative has identified a core group of activities, each already recommended by the participating agencies, that captures the best practices from programme experiences. Focusing initially on these activities will allow concerted action by the participating agencies and ensure consistent advice to country programmes and projects. Because of the focused and collaborative nature of this approach, it will increase the number of countries able to implement school health components of child-friendly school reforms and help to ensure that these programmes are scaled up. The core activities are seen as a starting point to which other interventions may be added, as appropriate.

The core activities also contribute to existing agency initiatives. They are an essential component of the "Health-Promoting Schools" initiative of WHO and of global efforts by UNICEF, UNESCO, and the World Bank to make schools effective as well as healthy, hygienic, and safe. Overall, the interagency action is perceived as Focusing Resources on Effective School Health and providing a FRESH Start to improving the quality and equity of education.

Focusing resources on the school-age child

A child's ability to attain her or his full potential is directly related to the synergistic effect of good health, good nutrition, and appropriate education. Good health and good education are not only ends in themselves but also means that provide individuals with the chance to lead productive and satisfying lives. School health is an investment in the future of a country and in the capacity of its people to thrive economically and as a society. An effective school health, hygiene, and nutrition programme offers many benefits, as outlined here.

• Responds to a new need

The success of child survival programmes and the greater efforts by many governments and communities to expand basic education coverage have resulted both in a greater number of school-age children and in a greater proportion of these children attending school. In many countries, targeted education programmes have ensured that many new school entrants are girls, for whom good health is especially important. Thus the school is now a key setting where the health and education sectors can jointly take action to improve and sustain the health, nutrition, and education of children who were previously beyond reach.

• Increases the efficacy of other investments in child development

School health programmes are the essential sequel and complement to early child care and development programmes. Increasing numbers of countries have programmes that ensure that a child enters school fit, well, and ready to learn, but the school-age child continues to be at risk of ill-health throughout the years of education. Continuing good health at school age is essential if children are to sustain the advantages of a healthy early childhood and take full advantage of what may be their only opportunity for formal learning. Furthermore, school health programmes can help ensure that children who enter school without benefit of early development programmes receive the attention they may need to take full advantage of their educational opportunity.

• Ensures better educational outcomes

Although schoolchildren have a lower mortality rate than infants, they do suffer from highly prevalent conditions that can adversely affect their development. Micronutrient deficiencies, common parasitic infections, poor vision and hearing, and disability can have a detrimental effect on school enrolment and attendance, and on cognition and educational achievement. In older children, avoidance of risky behaviours can reduce dropout due, for example, to early pregnancy. Ensuring good health at school age can boost school enrolment and attendance, reduce the need for repetition, and increase educational attainment, while good health practices can promote reproductive health and help avoid HIV/AIDS.

• Achieves greater social equity

As a result of universal strategies for basic education, some of the most disadvantaged children—girls, the rural poor, children with disabilities—now have access to school for the first time, but their ability to attend school, and to learn while there, is compromised by poor health. These are the children who will benefit most from health interventions, since they are likely to show the greatest improvements in attendance and learning achievement. School health programmes can thus help to modify the effects of socioeconomic and gender-related inequities.

• Is a highly cost-effective strategy

School health programmes help link the resources of the health, education, nutrition, and sanitation sectors in an infrastructure—the school—that is pervasive, sustained, and already in place. While the school system is rarely universal, coverage is often superior to that of health systems and has an extensive skilled workforce that already works closely with the community. The accessibility of school health programmes to a large proportion of each country's population, including staff as well as pupils, contributes to the low cost of programmes. The high effectiveness of these programmes is a consequence of the synergy between health benefits and the educational benefit. It is measurable in terms not only of improved health and nutrition but also of improved educational outcomes, reduced wastage, less repetition, and generally enhanced returns on educational investments.

The basic framework for an effective school health and nutrition programme

The framework described here is the starting point for developing an effective school health component in broader efforts to achieve more child-friendly schools. Much more could be done, but if all schools would implement the four intervention components discussed in the next section, there would be a significant immediate benefit and a basis for future expansion. In particular, the aim is to focus on interventions that promote learning through improved health and nutrition and that are feasible even in the most resource-poor schools and in hard-to-reach rural areas as well as in accessible urban areas. The interventions are known to be effective and are actively endorsed by all the supporting agencies: they constitute a framework from which individual countries can develop their own strategy to match local needs.

• Core framework for action: four components that should be made available together, in all schools

1. Health-related school policies

Health policies in schools, including skills-based health education and the provision of some health services, can help to promote the overall health, personal hygiene, and nutrition of children. However, good health policies should go beyond this, to ensure a safe and secure physical environment and a positive psychosocial environment; they should address issues such as abuse of pupils, sexual harassment, school violence, and bullying. By guaranteeing the further education of pregnant schoolgirls and young mothers, school health policies will help to promote inclusion and equity in the school environment. Moreover, policies that help to prevent harassment by other pupils, and even by teachers, also help to prevent girls from withdrawing or being withdrawn from schools. Policies regarding the health-related practices of teachers and pupils can reinforce health education: teachers can act as positive role models for their pupils, for example by not smoking in school. The process of developing and agreeing upon policies draws attention to these issues. The policies are best developed at many levels, including the national level; their development at the school level should involve teachers, children, and parents.

2. Provision of safe water and sanitation—the essential first steps towards a healthy learning environment

The school environment may damage the health and nutritional status of schoolchildren, particularly if it increases their exposure to hazards such as infectious disease carried by the water supply. Hygiene education is meaningless without clean water and adequate sanitation facilities. It is a realistic goal in most countries to ensure that all schools have access to clean water and adequate sanitation. By providing these facilities, schools can reinforce the health and hygiene messages and act as an example to both pupils and the wider community. This, in turn, can lead to a demand from the community for similar facilities. Sound construction policies will help to ensure that facilities address issues such as privacy and separate facilities for girls, particularly adolescent girls, which are important contributing factors to reducing dropout at menarche and even before.

3. Skills-based health education

A skills-based approach to health, hygiene, and nutrition education focuses on the development of the knowledge, attitudes, values, and life skills needed to make and act upon the most appropriate and positive health-related decisions. Health in this context extends beyond physical health to include psychosocial and environmental health issues. Changes in social and behavioural factors have given greater prominence to such health-related issues as HIV/AIDS, early pregnancy, injuries, violence, and tobacco and substance use. Unhealthy social and behavioural factors not only influence lifestyles, health, and nutrition but also hinder education opportunities for a growing number of school-age children and adolescents. The development of attitudes related to gender equity and respect between girls and boys, and of specific skills, such as dealing with peer pressure, are central to effective skills-based health education and to the creation of positive psychosocial environments. With appropriate attitudes and skills, individuals are more likely to adopt and sustain a healthy lifestyle during schooling and for the rest of their lives.

4. School-based health and nutrition services

Schools can effectively deliver certain health and nutritional services—provided that the services are simple, safe, and familiar—and address problems that are prevalent and recognized as important within the community. If these services can be provided by a school, the community will view the teacher and the school more positively, and teachers will perceive themselves as playing important roles. For example, micronutrient deficiencies and worm infections may be effectively dealt with by 6-monthly or annual oral treatment; changing the timing of meals or providing a snack to allay short-term hunger during school hours—an important constraint on learning—can contribute to school performance; and providing spectacles will allow some children to participate fully in class for the first time.

Supporting activities

The supporting activities outlined here provide the context in which the interventions can be implemented.

- *Effective partnerships between teachers and health workers and between education and health sectors*

The success of school health programmes demands an effective partnership between ministries of education and health and between teachers and health workers. The health sector retains responsibility for children's health but the education sector is responsible for implementing, and often funding, the school-based programmes. These sectors need to identify their various responsibilities and provide a coordinated approach to improving health and learning outcomes.

- *Effective community partnerships*

Promoting a positive interaction between the school and the community is fundamental to the success and sustainability of any school improvement process. Community partnerships engender a sense of collaboration, commitment, and communal ownership. They also build public awareness and strengthen demand. Within the school health component of such improvement processes, parental support and cooperation allow education about health to be shared and reinforced at home. The involvement of the broader community (the private sector, community organizations, and women's groups, for example) can enhance and reinforce school health promotion and resources. These partnerships, which should work together to make schools more child-friendly, can jointly identify health issues that need to be addressed through the school and then help design and manage appropriate activities.

- *Pupil awareness and participation*

Children must become important participants in all aspects of school health programmes and not simply the beneficiaries. They will learn about health in a practical sense by participating in health policy development and implementation; efforts to create a safer and more sanitary environment; health promotion aimed at their parents, other children, and community members; and school health services. This is an effective way of helping young people to acquire the knowledge, attitudes, values, and skills needed to adopt healthy lifestyles and to support health and education for all.

PART III
Anthelminthic chemotherapy

Anthelminthic chemotherapy

The control of disease due to helminth infections depends on the correct use of well tried, reliable prescriptions ranging from those based on medicinal plants (WHO, 1989) to those involving high-quality drugs developed by the research-based pharmaceutical industry (WHO, 1995, 1996, 1999). Details of the anthelminthic drugs recommended by the World Health Organization for the treatment of individuals in community programmes are set out in the current edition of *WHO model prescribing information: drugs used in parasitic diseases* (WHO, 1995).

Public health workers will be aware that generic forms of anthelminthic drugs are widely available. It is important to have confidence in the quality of these products before they are purchased and used; particular care should be taken to avoid counterfeit products. Anthelminthic drugs used in community programmes are more effective when they are applied against a background of the epidemiology of the infections of interest (Anderson & May, 1991; Albonico, Crompton & Savioli, 1998). There is always a risk of drug resistance emerging in a target population of helminths. Planning for helminth control programmes should not ignore this risk, and vigilance should be maintained to offset the emergence of anthelminthic drug resistance.

The Editors

References

Albonico M, Crompton DWT, Savioli L (1998) Control strategies for human intestinal nematode infections. *Advances in Parasitology*, 42:278–341.

Anderson RM, May RM (1991) *Infectious diseases of humans.* Oxford, Oxford University Press.

WHO (1989) *Medicinal plants in China.* Manila, WHO Regional Office for the Western Pacific.

WHO (1995) *WHO model prescribing information: drugs used in parasitic diseases*, 2nd ed. Geneva, World Health Organization.

WHO (1996) *Report of the WHO Informal Consultation on the Use of Chemotherapy for the Control of Morbidity due to Soil-Transmitted Nematodes in Humans.* Geneva, World Health Organization (document WHO/CTD/SIP/96.2).

WHO (1999) *Report of the WHO Informal Consultation on Monitoring of Drug Efficacy in the Control of Schistosomiasis and Intestinal Nematodes.* Geneva, World Health Organization (document WHO/CDS/CPC/SIP/99.1).

Treatment of soil-transmitted helminth infection: prescribing information for disease control

Marco Albonico[1]
Fondazione Ivo de Carneri
Milan, Italy

Introduction

The current aim of a helminth control programme is to lower the rates of morbidity and mortality. Control of disease due to helminthic infection is not achieved by a single approach but by a combination of actions aimed at preventing and reducing the infection and disease. One of the most effective strategies is the use of periodic chemotherapy (WHO, 1996a). The new challenges for the most appropriate use of anthelminthic drugs are: knowledge of the epidemiology in the region of intervention; using safe and effective drugs at reasonable costs; monitoring the impact of control activities; and ensuring the sustainability of control efforts.

In recent decades important progress has been made in anthelminthic chemotherapy. Effective, broad-spectrum, single-dose, safe, and relatively cheap drugs are now available for the treatment of helminth infections. In the 1990s, the idea of curing most of the major helminth infections (except dracunculiasis, trichinellosis, and fascioliasis) with the combined administration of three drugs—albendazole, ivermectin, and praziquantel—was appealing (Warren, 1990). Periodic chemotherapy has been recognized as one of the most cost-effective strategies in the short term for controlling morbidity due to soil-transmitted nematodes, including the negative effects of infection on growth, nutrition, and school performance, which are promptly reversed by treatment (Savioli, Bundy & Tomkins, 1992).

Although chemotherapy is the mainstay of most helminth control programmes, its rational use and the careful choice of the most appropriate drug will increase and prolong the benefits of treatment. The essential questions that public health planners should address when implementing a control programme are *why*, *whom*, *when*, and *how* to treat. Parasitological, environmental, and socioeconomic variables mean that different answers will be found in each endemic country. Indeed, control measures should not normally be introduced without reliable knowledge of the current epidemiological situation. Conditions might be expected to vary

[1] Dr Marco Albonico, Fondazione Ivo de Carneri, Via Quattro Marzo 14, 10122 Torino, Italy
Tel: +39 11 431 0218, Fax: +39 11 436 1474, E-mail: albonico@tin.it

quite markedly even within the same climatic, topographical, and socioeconomic region. The epidemiological baseline must therefore be established before control measures are introduced.

Why treat?

In the short term, treatment is intended to prevent and control morbidity in infected populations by reducing the worm burden. In the long term, its purpose is to prevent healthy people from becoming infected by reducing the source of infection, and to promote health education and integration of control activities with other programmes at the primary health care level.

The aim of periodic chemotherapy, even when there will be reinfection, is to keep the intensity of infection below the threshold that can cause disease. Periodic chemotherapy not only cures the acute symptoms but may also reverse the negative effects of chronic infections on nutrition, growth, and children's school performance (WHO, 1996a).

Whom to treat?

Universal—Everybody is treated. When resources are available, this approach is recommended for any overall prevalence of infection and a prevalence of heavy infections that exceeds 10%.

Targeted—Certain groups are targeted for treatment. School-age children, preschool children, and women of childbearing age are more prone to the negative consequences of disease. This approach is recommended when there is high prevalence of infection (>50%) but prevalence of heavy infections is less than 10%.

Selective—Infected subjects are treated. As a rule, passive case management at the peripheral health unit is the recommended approach where there is low prevalence of infections and of heavy infections (<50% and <10% respectively) (Montresor et al., 1998).

When (how often) to treat?

The appropriate frequency for chemotherapy depends on the local epidemiological situation and on the efficacy of the drug concerned. A model proposed by Anderson & May (1985) for control of *Ascaris lumbricoides* transmission implies treatment of 31% of the community every 2 months with a drug that is 90% effective. In deciding when and how frequently to treat a particular population to control disease, it is essential both to assess the magnitude of transmission (rate of reinfection following chemotherapy) and to identify seasonal influences on transmission of infective stages. Seasonal effects can be quite subtle, being related, for example, to the rainy season or to such activities as agricultural fertilization using nightsoil. Information produced by the Asian Parasite Control Organization (APCO) has shown quite clearly the importance of such factors on the transmission of *Ascaris lumbricoides* (Kobayashi, 1980). Seasonality is also important in monitoring the impact of chemotherapy on transmission of hookworms, as shown by a recent study in Zanzibar (Albonico et al., 1999; see Figure 1).

The appropriate frequency of community treatment should therefore be determined after consideration of epidemiological, pharmacological, socioeconomic, and environmental factors. From the epidemiological viewpoint, the key issues are the frequency and seasonal variation of reinfection. Ideally, community treatment should be applied soon after the end of an identified transmission season.

Results obtained by the APCO workers indicate that the appropriate frequency for anthelminthic chemotherapy to control morbidity induced by soil-transmitted nematodes is twice a year for prevalence below 50% and three times a year where prevalence exceeds 50%. Again, it is stressed that the decision about intervals between treatments must be taken after consideration of local circumstances (WHO, 1996a).

Cost-effectiveness also plays a major role in determining the frequency of treatment, although studies of this factor have been carried out only for roundworms. The most cost-effective intervention is the yearly treatment of individuals in areas of high transmission, which immediately and dramatically reduces

Figure 1. Intensity of intestinal helminthiases in the untreated cohort (left, $n = 1037$) and in the treated cohort (right, $n = 1011$) given mebendazole, 500 mg every four months (time 0, 4, 8), Pemba Island, 1995[a]

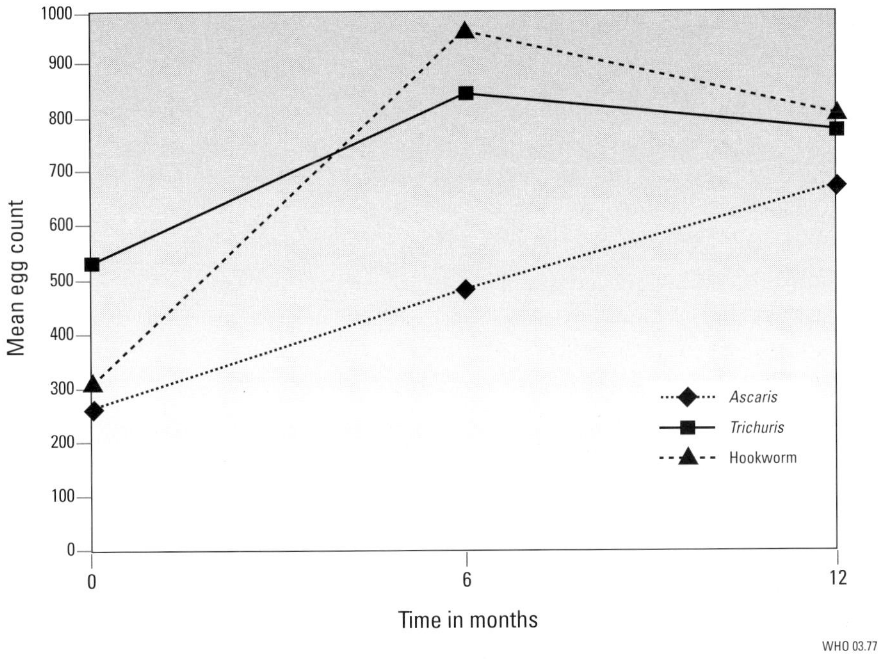

[a] Albonico M et al., unpublished observations.

morbidity. Subsequent doses within the same year produce less impressive results but maintain morbidity at low levels (Guyatt, Bundy & Evans, 1993). Where financial resources are limited, it is more efficient to treat a large proportion of the population less frequently than to treat a small proportion more often (Guyatt et al., 1995). Mass and targeted approaches have been shown to be considerably more cost-effective than the selective approach (Holland et al., 1996).

How to treat?

Five anthelminthic drugs described in the WHO Model List of Essential Medicines may be considered for the single-dose treatment of soil-transmitted helminths and of filariae: albendazole, mebendazole, pyrantel, levamisole, and ivermectin (WHO, 2003).

Albendazole 400 mg

Chewable tablets 200 mg and 400 mg.
Benzimidazole derivative.
Mode of action. Albendazole is absorbed by the intestinal cells of the worms; by binding to helminth tubulin, it blocks glucose uptake and inhibits the formation of ATP, which is required by the worms for survival and reproduction (Lacey, 1990).

Levamisole 2.5 mg/kg

Tablets 40 mg, syrup 40 mg/5 ml.
Laevorotatory isomer of tetramisole.
Mode of action. Binds to acetylcholine receptors and inhibits the production of succinate dehydrogenase, causing spastic paralysis and passive elimination of the worms (Martin, 1993).

Mebendazole 500 mg

Chewable tablets 100 mg and 500 mg, suspension 100 mg/5 ml.
Benzimidazole derivative.
Mode of action. As for albendazole.

Pyrantel 10 mg/kg

Chewable tablets 250 mg, suspension 50 mg/ml.
Pyrimidine derivative.
Mode of action. Binds to acetylcholine receptors and paralyses the worms by depolarizing the neuromuscular junctions (Aubry et al., 1970).

Ivermectin 200 μg/kg

Chewable tablets 6 mg.
Macrocyclic lactone.

Mode of action. Causes paralysis in many nematodes and arthropods through the influx of chloride ions across cell membranes and the disruption of neural transmission mediated by γ-aminobutyric acid (GABA) (Abalis, Eldefrawi & Eldefrawi, 1986).

Anthelminthic activity of these drugs is summarized in Table 1.

Rationale for the choice and appropriate use of drugs

The choice of appropriate anthelminthic drug for use in a control programme depends on safety, therapeutic effect (efficacy), spectrum of activity, local health policy, and financial considerations. Other important issues to consider are the accessibility—including drug delivery—and acceptability of treatment, and possible integration with other control programmes. The possibility of drug resistance should also be carefully monitored when periodic chemotherapy is considered. With these factors in mind, the choice of anthelminthic drug for public health use should be tailored to the local epidemiology of soil-transmitted helminth infections. Information should be collected on prevalence and intensity of helminth infections, on population groups at highest risk of morbidity, and on the health impact of helminth infections in the community to be treated. This information should be used as the rationale for local health planners to choose the most appropriate anthelminthic drug and should be the backbone of a successful chemotherapy-based control programme (Albonico, Crompton & Savioli, 1999).

Baseline epidemiology

Baseline data on the local epidemiology of intestinal nematodes infection are essential for decisions on the most appropriate use of periodic treatment. Pilot studies should be performed to assess:

- which species of helminths are prevalent in the community (including distinction between *Ancylostoma duodenale* and *Necator americanus*);
- the intensity of these infections (prevalence of heavy infections);
- the extent of transmission (reinfection rate);
- whether there is a seasonality of transmission;
- which population groups are most at risk of morbidity; and
- the health impact of these infections in the community (association with anaemia, diarrhoea, malnutrition, impaired physical and cognitive development).

Safety information

Overall, the side-effects of anthelminthic drugs are negligible and important adverse events are virtually non-existent (see Table 2). However, the safety of anthelminthic drugs is of utmost importance because large portions of the

Table 1. Recommended drugs for treatment of soil-transmitted nematode infections

Anthelminthic	Therapeutic activity		Dosage	Use in pregnancy and in children
Albendazole	Ascariasis		400 mg single dose	Not recommended in the first trimester of
(tablets 200 and 400 mg,	+++		400 mg single dose	pregnancy or in children under 2 years
suspension 100 mg/5 ml)	Trichuriasis	++	400 mg single dose	
	Hookworm infections		400 mg daily for 3 days	
	+++			
	Strongyloidiasis	++		
Ivermectin	Ascariasis		200 µg/kg single dose	Not recommended in the first trimester of
(tablet 6 mg)	+++		200 µg/kg single dose	pregnancy
	Trichuriasis	+		
	Hookworm infections	—	200 µg/kg single dose	
	Strongyloidiasis			
	+++			
Levamisole	Ascariasis		2.5 mg/kg single dose	No evidence of teratogenicity. Not
(tablet 40 mg,	+++		2.5 mg/kg single dose	recommended in the first trimester of
syrup 40 mg/5 ml)	Trichuriasis	+	2.5 mg/kg single dose. For heavy *Necator*	pregnancy
	Hookworm infections	++	infection, repeat after 7 days	
	Strongyloidiasis	—		
Mebendazole	Ascariasis		500 mg single dose	Not recommended in the first trimester of
(tablets 100 and 500 mg,	+++		100 mg twice daily for 3 days, 500 mg	pregnancy or in children under 1 year
suspension 100 mg/5 ml)	Trichuriasis	++	single dose (less effective)	
			100 mg twice daily for 3 days, 500 mg	
	Hookworm infections		single dose (less effective)	
	+++		100 mg twice daily for 28 days	
	Strongyloidiasis	+		
Pyrantel	Ascariasis		10 mg/kg single dose	Not recommended in the first trimester of
(tablets 250 mg,	+++			pregnancy
suspension 50 mg/ml)	Trichuriasis	—	10 mg/kg single dose. For heavy *Necator*	
	Hookworm infections	++	infection, repeat for 3 days	
	Strongyloidiasis	—		

Table 2. Side-effects of the most commonly used anthelminthic drugs

Anthelminthic	Occasional (1–5%)	Rare (<1%)	Rare (<1%) in high dosages
Albendazole		*GI symptoms:* epigastric pain, diarrhoea, nausea, vomiting *CNS symptoms:* headache, dizziness *Allergic phenomena:* oedema (0.7/1000), rashes (0.2/1000), urticaria (0.1/1000)	*Liver abnormalities:* raised transaminase *Bone marrow depression:* leucopenia
Ivermectin	*GI symptoms:* borborygmus, diarrhoea, constipation *CNS symptoms:* headache		
Levamisole	*GI symptoms:* vomiting *CNS symptoms:* dizziness, headache, weakness		
Mebendazole		*GI symptoms:* transient abdominal pain, diarrhoea	*GI symptoms:* severe abdominal pain *Liver abnormalities:* raised transaminase *Allergic phenomena:* skin rashes *CNS symptoms:* vertigo, headache. *Bone marrow depression:* neutropenia
Pyrantel	*GI symptoms:* diarrhoea, abdominal pain, nausea, vomiting *CNS symptoms:* headache *Liver abnormalities:* raised transaminase		

population are normally treated; if there are contraindications to use of the drug, significant risk groups may be excluded from treatment.

For pregnant women in areas where hookworm infections are endemic (prevalence >20–30%) and anaemia is prevalent, WHO recommends treatment with albendazole, levamisole, mebendazole, or pyrantel. These drugs can be given to pregnant and lactating women but should be avoided in the first trimester of pregnancy (WHO, 1996b).

The use of anthelminthics in children under 2 years of age has been constrained by two factors: statements by manufacturers that benzimidazole anthelminthics are contraindicated in that age group, and the assumption that young children do not suffer important morbidity due to helminthic infections. However, a number of studies have shown the absence of side-effects in this age group (Mbendi et al., 1988; Albonico et al., 1994a). Moreoever, when mebendazole, 500 mg, was used in a randomized trial in children aged 6–72 months in Zanzibar, the reported side-effects were negligible. Mebendazole was also shown to have a beneficial effect in children under the age of 2 years in Zanzibar, reducing severe anaemia and wasting malnutrition and improving motor development (R.J. Stoltzfus, personal communication). These and similar observations imply the need for a reappraisal of the use of anthelminthic drugs in young children living in endemic areas.

Effectiveness

In planning control programmes, the effectiveness of an anthelminthic (the effect of the drug against an infective agent in a particular host, living in a particular environment and interacting with ecological, immunological, and epidemiological determinants) is a more important consideration than its efficacy (the effect of the drug against an infective agent in ideal experimental conditions rather than in a specific context). However, evaluation of the effectiveness of a drug in bringing about a general improvement in health status by reducing morbidity and mortality should also take account of factors such as the reliability of drug delivery, the accessibility and acceptability of treatment, and the costs of sustainability. Assessing the impact of treatment requires identification of the most useful morbidity indicators to define the negative health consequences of helminth infections.

Classically, most studies have evaluated the impact of treatment on direct or indirect measurements of helminth infections, such as reduction of prevalence and intensity of infection, the number of adult worms expelled, or the reduction in the number of eggs passed in faeces. More recently, attention has focused more on other effects, such as the impact on nutrition, and the improvement in iron status and cognitive function of infected children (Bundy et al., 1992).

The advantages of using prevalence and intensity of infection as parasitological indicators are that they are easy to measure and that standardized methods are available for their measurement. The disadvantages are that these indicators do not always reflect morbidity, and that there is variability related to the study population, to the intensity of infection (and of transmission), and to different epidemiological settings.

The advantages of nutritional and psychometric indicators lie in their more relevant and direct correlation with morbidity, and in revealing health and socioeconomic results that are more likely to mobilize resources from ministries of health and funds from donor agencies. Disadvantages are the greater difficulty of assessing these indicators, the lack of standardized methods, and the wide variability due to confounding factors and associated causes that may jeopardize the reliability of data.

Measuring the effects of anthelminthic drugs

Efficacy is usually measured by assessing, in qualitative and quantitative diagnostic tests, the eggs or larvae in faeces after an optimal time interval (which varies with the parasite). Cure rate (CR) and egg reduction rate (ERR) are indicators commonly used to measure the reduction in prevalence and reduction in intensity of infection. However, drug efficacy varies widely, even in trials in which the same drug is given at the same dosage (Table 3). This may be explained by the failure to use a standardized parasitological technique, the variable interval before post-treatment follow-up, the presence of single or multiple infections before treatment, the different pretreatment prevalence and intensity rates, the strain susceptibility, the age group enrolled in the study, the drug quality, the sample size, and the fact that many trials are not placebo-controlled. In addition, diverse statistical measurements of intensity (arithmetic or geometric means) have been used, making the results of efficacy trials more difficult to compare and to interpret. The standardized guidelines recommended by WHO (WHO, 1999) should allow the results of surveys to be compared.

The various hypotheses that may account for "poor" efficacy of anthelminthic drugs and that should be taken into account may be summarized as follows:

— poor drug quality
— reduced absorption and bioavailability
— heavy intensity of infection
— variability of egg production and excretion
— heavy transmission of infections (immature worms)
— parasitological examination performed too soon after treatment
— poor strain susceptibility (tolerance)
— induced drug resistance.

Drug delivery

The delivery of treatment is another variable to consider in evaluating the use of chemotherapy. Experience from helminth control programmes (Renganathan et

Table 3. Review of some comparative trials of single-dose anthelminthic drugs

Study	Drug	No. of patients	*Ascaris*		*Trichuris*		Hookworms		Technique and post-treatment examination
			CR%	ERR%	CR%	ERR%	CR%	ERR%	
Moens et al., 1978,	Lev	768	91	98					Kato–Katz
multi-centre study	Pyr	157	87	96					30 days
Kan, 1986, Malaysia	Alb	33			33[a]	88			Kato–Katz
	Pyr	51			68	90			30–45 days
Mbendi et al., 1985, Zaire	Alb	450	100		100		100		Formol ether
	Pyr	125	100		0		88		7 days
Sinniah, Chew & Subramaniam,	Alb	50	91	99	42	71	100	100	Flotation + Beavers
1990, Malaysia	Pyr	50	90	97	56	60	43	64	21 days
Ismail, Premaratne & Suraweera,	Alb	160	96	100	32	87	100[b]	100	Kato–Katz
1991, Sri Lanka	Lev	150	86	96	18	73	87[b]	100	14–28 days
	Meb	181	97	99	36	80	90[b]	100	
	Pyr	162	94	99	23	54	90[b]	100	
Jongsuksuntingul et al., 1991,	Alb	53	100	100	67	87	84	96	Kato–Katz
Thailand	Meb	56	100	100	70	89	30	70	14 days
Long-Qi et al., 1992, China	Alb	100	99	100	42	69	95	99	Kato–Katz
	Pyr	50	81	82	7	17	83	90	28 days
Bartoloni et al., 1993, Bolivia	Alb	54	100	100	33[b]	46	82	93	Kato–Katz
	Meb	54	100	100	60[b]	15	17	62	21–28 days
Albonico et al., 1994b, Zanzibar	Alb	1174	99	100	10	73	57	98	Kato–Katz
	Meb	1120	98	99	14	82	22	82	21 days
De Clercq et al., 1997, Mali	Meb	35					23	0	Kato–Katz
	Pyr	29					45	75	28 days

al., 1995; Montresor et al., 1997) and from the work of the Partnership for Child Development (WHO, 1994) has shown that coverage of children in school is very effective. Moreover, treatment targeted at school-age children can reduce the intensity of infection in the remainder of the community who receive no treatment, indicating that treating school-age children reduces transmission in low-transmission areas (Bundy et al., 1990; Asaolu, Holland & Crompton, 1991). Reaching children through schools has several advantages:

— large numbers of children can be reached
— treatment can be administered by teachers
— other conditions such as schistosomiasis and micronutrient deficiencies can be treated/prevented
— health education can be provided and parents can be involved.

However, a large number of children of school age are not enrolled, and ways to reach this neglected group are lacking. A recent study carry out in the United Republic of Tanzania has shown that schools can be an efficient tool to reach the non-enrolled. The child-to-child communication approach has proved to be an excellent strategy, with a coverage of more than 90% of non-enrolled children (Montresor, personal communication). Anthelminthic treatment can also be delivered through maternal and child health clinics, thus reaching both mothers and their children.

Accessibility to treatment is a major constraint in countries where most people live far from schools and health centres. The community can be treated in active campaigns integrated with the filariasis elimination programme or in a self-treatment approach as used for chemotherapy for onchocerciasis (WHO, 1996a).

Acceptability and ancillary benefits of chemotherapy

The acceptability of treatment, taking account of side-effects and ancillary benefits, is important in chemotherapy of helminth infections. Treatment is usually well accepted, both because side-effects are negligible and because of the immediate benefits perceived by the community. Helminth control programmes have therefore been used as entry points to promote other health issues, such as family planning (Yokogawa, 1985), health education, and latrine construction and maintenance (Albonico et al., 1997).

Costs

Drug costs are an essential consideration when dealing with a large number of treatments in countries with limited health budgets. The total cost of treatment, including such factors as storage, delivery to site, on-site treatment, and monitoring, should be estimated (WHO, 1997a) and carefully balanced against the benefits. Cost-effectiveness analysis is a complex but essential part of planning that guides the choice of drug and the frequency of treatment. The cost-effectiveness measure for disease control can be the average mean worm burden prevented per person, or the amount of morbidity prevented per person (in terms of blood loss, vitamin A deficiency, growth retardation, etc.). Alternatively, the cost per person treated can be estimated (Holland et al., 1996). Integration with other control programmes provides a means of sharing costs and improving benefits. For example, control of schistosomiasis and control of lymphatic filariasis require a similar approach; the strategy recommended by WHO (WHO, 1997) is shown in Table 4.

Drug costs will vary depending on the health policy of the country concerned and the quantity of drugs required, and on whether generic products are produced locally. However, the cost of the single-dose anthelminthics averages a few US cents per dose.

Quality

The cheapest drug is not always the best; quality is of paramount importance. In different preparations there may be variations in pharmaceutical formulation, in bioavailability, and in therapeutic equivalence. Before purchasing any anthelminthic drug, especially a generic product, programme managers must take steps to avoid fake or counterfeit drugs. If drugs are not purchased directly from a research-based pharmaceutical company, which will invariably guarantee the quality, quality must be assured by independent analysis: proof is required that a product is not counterfeit or substandard and that it actually contains the stated compound. A simple thin-layer chromatographic method has been produced for the identification of impurities and active ingredients in essential drugs such as albendazole and mebendazole (Pachaly et al., 1994). It is also important to know such details as the quantity of active compounds and excipients used, and the results of disintegration and dissolution tests using standard pharmacopoeial methods. Contracts for the purchase of anthelminthic drugs should be dependent on adequate quality being assured by independent analysis.

This issue was stressed in a workshop held in the Seychelles on the control of intestinal nematodes in countries of the Indian Ocean area (WHO, 1996c) and raised again by WHO in recent informal consultations (WHO, 1996a, 1999).

Drug resistance

The possibility of the emergence of drug resistance should be considered when the use of anthelminthic chemotherapy is evaluated. Increasing resistance to anthelminthics is well documented in animal nematodes (Geerts, Coles & Gryseels, 1997). The molecular basis of benzimidazole resistance has been identified for animal nematodes as a point mutation of the amino acid 200 in the gene that codes for β-tubulin, the site of action of benzimidazoles (Kwa, Jetty & Roos, 1994). There is concern that frequent use of chemotherapy to control human

Table 4. Strategy recommended by WHO for the integrated control of filariasis, intestinal helminth infection, and schistosomiasis[a]

Infection	Target group	Intervention	Treatment	Frequency
Schistosomiasis	School-age children	Mass treatment in schools where prevalence of haematuria is >0%; selective treatment in schools where prevalence of haematuria is <20%	Praziquantel	Every 12 months
Intestinal helminths	Preschool children	Treatment in nurseries or in MCH clinics	Albendazole or levamisole or mebendazole or pyrantel	Every 4–12 months (depending on transmission)
	School-age children	Mass treatment in schools		Every 4–12 months (depending on transmission)
	Pregnant women	Treatment in MCH clinics		Once after the first trimester
Filariasis	Whole population	Community treatment except in pregnant women and in children under 2 years	Ivermectin and albendazole	Every 12 months
		Disease management in patients with elephantiasis	Nursery care and treatment of disabilities	

[a] Adapted from WHO, 1997.

helminthiasis may select resistant worms and thus reduce the benefits of treatment at both individual and public health levels. As yet, however, field data on the occurrence of resistance in human helminths are very limited. There have been recent reports from small-scale studies of failure in the treatment of human hookworm infections following frequent usage (not mass or targeted treatment) of mebendazole in Mali (De Clercq et al., 1997) and of pyrantel in north-west Australia (Reynoldson et al., 1997).

Predisposing factors for resistance

A theoretical analysis of factors in community-based treatment that predispose to the development of drug resistance in helminths suggests that the following may have an influence:

- isolation of helminth populations to the extent that there is no gene flow between treated and untreated populations
- universal anthelminthic treatment (i.e. treatment of the whole population)
- treatment is homogeneous (i.e. uses the same drug)
- continual selection pressure is applied (i.e. frequent treatments, at intervals shorter than the generation time of the helminths, so that only resistant worms contribute to ongoing transmission)
- at the individual level, under-dosing or treatment with only partially effective drugs (WHO, 1999).

Under these conditions, it would appear to be inevitable that drug resistance will appear in soil-transmitted helminths. The genes involved already exist in the different helminth populations, so that the question becomes one of when resistant strains will be selected and become a significant proportion of the total population.

A key factor influencing the spread of a drug-resistant genotype in a population is the percentage of helminths in those not exposed to the drug, which will determine the proportion of worms surviving to therapy in the next generation of worms. Computer models of worm populations and the evolution of anthelminthic resistance in animal helminths suggest that development of resistance will be delayed if only 80% of the infected population is treated, (Barnes, Dobson & Barger, 1995). These models also indicate that development of resistance is slower when two drugs with different modes of action are used simultaneously.

The appearance of resistance will also be influenced by the timing of treatment in relation to the transmission season. A survey of anthelminthic resistance in nematodes of sheep and goats in Greece has shown that benzimidazole resistance has developed on some small islands that suffer from extended droughts. Treatment during dry periods will select strongly for resistance, because very few or no helminth stages survive outside the host (Coles, 1995).

Monitoring and management of emerging anthelminthic drug resistance in humans

The fundamental strategy is to apply chemotherapy so that the emergence of drug resistance is delayed or prevented while health benefits continue to accrue. Several measures can thus be taken to delay and even avoid the problem of drug resistance in human helminths. For example, treating only a proportion of the people in an infected population (e.g. by targeting school-age or preschool children, or women of childbearing age) will ensure that a proportion of the helminth population remains untreated to serve as a reservoir of drug-susceptible alleles, thereby diluting the selection pressure. Treatment at intervals greater than the helminth generation time will act against drug resistance. Haemonchosis in sheep is treated 10–15 times per year, but such frequent treatment would never be used in human helminth control programmes. Use of combinations of anthelminthic drugs in control programmes also tends to reduce and delay the degree of selection pressure on the population of helminths (Savioli et al., 1997); computer models indicate that drug combinations are more effective in delaying resistance than alternating use of different drugs (Barnes, Dobson & Barger, 1995).

Since it is not known whether current treatment protocols are likely to lead to drug resistance, it is essential that measures be put in place to recognize and monitor resistance. At present, only comparison of reductions in intensity (egg reduction rate) following treatment with efficacy of the drug when used for the first time) provides data that would suggest the appearance of anthelminthic drug resistance in humans. A standard protocol for monitoring drug effectiveness in vivo has been suggested by WHO (WHO, 1999).

There is a need for reliable and valid in vitro tests to confirm suspected drug resistance in humans under the conditions prevailing in developing countries. The Egg Hatch Assay test, used in veterinary medicine to assess benzimidazole resistance in field conditions (Coles et al., 1992), is being evaluated in humans. Its repeatability and standardization are under study but results are promising (Albonico, personal communication). However, with the current in vivo and in vitro techniques, drug resistance will be detected only when 25% or more of the worm population is already resistant—a condition that is unlikely to be reversible (Martin, Anderson & Jarret, 1989).

For research purposes, attempts should be made to isolate suspected resistant worms from the field for study and, where possible, to maintain them under laboratory conditions. Polymerase chain reaction-based assays are being used to detect mutations in receptor genes that are relevant for drug action, in order to develop a genotypic test to identify the resistant genotype before it becomes clinically apparent.

Conclusions

The feasible yet challenging goal of controlling disease due to helminth infection has been widely endorsed. Chemotherapy is essential to achieve this goal, although it is not a "magic bullet" and should be used according to rational strategies. Recommendations on the best use of anthelminthic drugs are not limited to the prescribing information provided by the drug manufacturers, nor are they all included in this brief paper. Health managers in endemic countries need to adapt use of anthelminthics to the local epidemiological and socioeconomic conditions. Once a rational choice of drug has been made, appropriate use of the drug in a chemotherapy-based control programme can be assessed only by a reliable and effective monitoring system. Finally, analysis of cost-effectiveness is essential, always bearing in mind the cost of having no helminth control programme.

References

Abalis IM, Eldefrawi AT, Eldefrawi ME (1986) Actions of avermectin B1a on the gamma-aminobutyric acid A receptor and chloride channels in rat brain. *Journal of Biochemical Toxicology*, 1:69–82.

Albonico M, Crompton DWT, Savioli L (1999) Control strategies for human intestinal nematode infections. *Advances in Parasitology*, 42:277–341.

Albonico M et al. (1994a) Efficacy of a single dose of mebendazole on prevalence and intensity of soil-transmitted nematodes in Zanzibar. *Tropical and Geographical Medicine*, 46:142–146.

Albonico M et al. (1994b) A randomised controlled trial comparing mebendazole and albendazole against *Ascaris*, *Trichuris*, and the hookworms. *Transactions of the Royal Society of Tropical Medicine and Hygiene*, 88:585–589.

Albonico M et al. (1997) Fewer parasites in paradise. Achievements and future perspectives of the control of intestinal parasitic infections in the Seychelles. *Seychelles Medical and Dental Journal*, 5:3–7.

Albonico M et al. (1999) A controlled evaluation of two school-based antihelmintic chemotherapy regimens on intensity of helminth infections. *International Journal of Epidemiology*, 28:591–596.

Anderson RM, May RM (1985) Helminth infections of humans: mathematical models, population dynamics and control. *Advances in Parasitology*, 24:1–101.

Aubry ML et al. (1970) Aspects of the pharmacology of a new anthelmintic: pyrantel. *British Journal of Pharmacology*, 38:332–344.

Barnes EH, Dobson RJ, Barger IA (1995) Worm control and anthelmintic resistance: adventures with a model. *Parasitology Today*, 11:56–63.

Bartoloni A et al. (1993) Comparative efficacy of a single 400 mg dose of albendazole or mebendazole in treatment of nematode infections in children. *Tropical and Geographical Medicine*, 45:114–116.

Bundy DAP et al. (1990) Control of geohelminths by delivery of targeted chemotherapy through schools. *Transactions of the Royal Society of Tropical Medicine and Hygiene*, 84:115–120.

Bundy DAP et al. (1992) Evaluating measures to control intestinal parasitic infections. *World Health Statistics Quarterly*, 45:168–179.

Coles GC (1995) Chemotherapy of human nematodes: learning from the problems in sheep. *Journal of the Royal Society of Medicine*, 88:649P–651P.

Coles GC et al. (1992) World Association for the Advancement of Veterinary Parasitology (WAAVP) methods for the detection of anthelmintic resistance in nematodes of veterinary importance. *Veterinary Parasitology*, 44:35–44.

De Clercq D et al. (1997) Failure of mebendazole in treatment of human hookworm infections in the southern region of Mali. *American Journal of Tropical Medicine and Hygiene*, 57:25–30.

Geerts S, Coles GC, Gryseels B (1997) Anthelmintic resistance in human helminths: learning from the problem with worm control in livestock. *Parasitology Today*, 13:149–151.

Guyatt HL, Bundy DAP, Evans D (1993) A population dynamic approach to the cost-effectiveness analysis of mass anthelminthic treatment: effects of treatment frequency on *Ascaris* infection. *Transactions of the Royal Society of Tropical Medicine and Hygiene*, 87:570–575.

Guyatt HL et al. (1995) Control of *Ascaris* infection by chemotherapy: which is the most cost-effective option? *Transactions of the Royal Society of Tropical Medicine and Hygiene*, 89:16–20.

Holland CV et al. (1996) A cost-effective analysis of anthelminthic intervention for community control of soil-transmitted infection: levamisole and *Ascaris lumbricoides*. *Journal of Parasitology*, 82:527–530.

Ismail MM, Premaratne UN, Suraweera MGW (1991) Comparative efficacy of single dose anthelminthics in relation to intensity of geohelminth infections. *Ceylon Medical Journal*, 36:162–167.

Jongsuksuntigul P et al. (1991) Comparative study of albendazole and mebendazole in treatment of ascariasis, hookworm infections and trichuriasis. *Southeast Asian Journal of Tropical Medicine and Public Health*, 24:724–729.

Kan SP (1986) The efficacy of some broad spectrum anthelminthics in the treatment of *Trichuris trichiura* infections. In: Yokogawa M, ed. *Collected papers on the control of soil-transmitted helminthiases, Vol. 3.* Tokyo, Asian Parasite Control Organization:71–76.

Kobayashi A (1980) Studies on the mode of *Ascaris* infection in Japan. In: Yokogawa M, ed. *Collected papers on the control of soil-transmitted helminthiases, Vol. 1.* Tokyo, Asian Parasite Control Organization:109–118.

Kwa MSG, Jetty GV, Roos MH (1994) Benzimidazole resistance in *Haemoncus contortus* is correlated with a conserved mutation at aminoacid 200 in β tubulin isotype 1. *Molecular and Biochemical Parasitology*, 63:299–303.

Lacey E (1990) Mode of action of benzimidazoles. *Parasitology Today*, 6:107–112.

Long Qi X et al. (1992) Treatment of soil-transmitted helminth infections by anthelminthics in current use. *Chinese Journal of Parasitology and Parasitic Diseases*, 10:95–99.

Martin PJ, Anderson N, Jarret RG (1989) Detecting benzimidazole resistance with faecal egg count reduction tests and in vitro assays. *Australian Veterinary Journal*, 66:236–240.

Martin RJ (1993) Neuromuscular transmission in nematodes parasites and antinematodal drug action. *Pharmacology and Therapeutics*, 58:13–50.

Mbendi N et al. (1985) Albendazole in the treatment of intestinal nematode infections in Kinshasa, Zaire. *Annales de la Société Belge de Médecine Tropicale*, 65:41–47.

Mbendi M et al. (1988) L'albendazole dans le traitement des nematodes intestinales chez l'enfant agé de 1 à 2 ans. [Albendazole in the treatment of intestinal nematodes in children aged 1–2 years.] *Médecine et Chirurgie Digestives*, 17:213–215.

Moens M et al. (1978) Levamisole in ascariasis: a multicentre controlled evaluation. *American Journal of Tropical Medicine and Hygiene*, 27:897–904.

Montresor A et al. (1997) Enquête préliminaire à la mise en place d'un programme de santé scolaire en Guinée. [Preliminary study of the implementation of a school health study in Guinea.] *Médicine Tropicale*, 57:294–298.

Montresor A et al. (1998) *Guidelines for the evaluation of soil-transmitted helminthiasis and schistosomiasis at the community level.* Geneva, World Health Organization (document WHO/CTD/SIP.98.1).

Pachaly P et al. (1994) *Simple thin-layer chromatographic identification of active ingredients in essential drugs.* Aulendorf, Germany, German Pharma Health Fund.

Renganathan E et al. (1995) Evolution of operational research and development of a national control strategy against intestinal helminths in Pemba Island, 1988–92. *Bulletin of the World Health Organization*, 73:183–190.

Reynoldson JA et al. (1997) Failure of pyrantel in treatment of human hookworm infections (*Ancylostoma duodenale*) in the Kimberly region of North West Australia. *Acta Tropica*, 68:301–312.

Savioli L, Bundy DAP, Tomkins A (1992) Intestinal parasitic infections: a soluble public health problem. *Transactions of the Royal Society of Tropical Medicine and Hygiene*, 86:353–354.

Savioli L et al. (1997) Anthelminthic resistance in human helminths: learning from the problem with worm control in livestock—Reply. *Parasitology Today*, 13:156.

Sinniah B, Chew PI, Subramaniam K (1990) A comparative trial of albendazole, mebendazole, pyrantel pamoate and oxantel pyrantel pamoate against soil-transmitted helminthiases in school children. *Tropical Biomedicine*, 7:129–134.

Warren KS (1990) An integrated system for the control of the major human helminth parasites. *Acta Leiden*, 59:433–442.

WHO (1994) *Meeting between the World Health Organization and the Partnership for Child Development.* Geneva, World Health Organization (document WHO/CDS/IPI/94.1).

WHO (1996a) *Report of the WHO Informal Consultation on the Use of Chemotherapy for the Control of Morbidity due to Soil-Transmitted Nematodes in Humans, Geneva, 29 April to 1 May 1996.* Geneva, World Health Organization (document WHO/CTD/SIP.96.2).

WHO (1996b) *Report of the WHO Informal Consultation on Hookworm Infection and Anaemia in Girls and Women, Geneva, 5–7 December 1994.* Geneva, World Health Organization (document WHO/CTD/SIP.96.1).

WHO (1996c) *Report of the WHO Workshop on the Control of Intestinal Parasitic Infections in the Seychelles and in Neighbouring Countries, Mahé, 23–24 October 1995.* Geneva, World Health Organization (document WHO/CTD/SIP.96.5).

WHO (1997) *Division of Control of Tropical Diseases progress report 1996.* Geneva, World Health Organization (document WHO/CTD/PR/97.1).

WHO (1999) *Report of the WHO Informal Consultation on Monitoring of Drug Efficacy in the Control of Schistosomiasis and Intestinal Nematodes, Geneva, 8–10 July 1998.* Geneva, World Health Organization (document WHO/CDS/CPC/SIP/99.1).

WHO (2003) *Selection and use of essential medicines. Report of the WHO Expert Committee (including the 12th Model List of Essential Medicines).* Geneva, World Health Organization, (WHO Technical Report Series, No. 914).

Yokogawa M (1985) JOICFP'S experience in the control of ascariasis within an integrated programme. In: Crompton DWT, Nesheim MC, Pawlowski ZS, eds. *Ascariasis and its public health significance.* London, Taylor & Francis:265–277.

Control of intestinal helminthiasis in pregnancy—the Sri Lankan experience

M.M. Ismail,[1] T.M.S. Atukorala, T.S. Naotunne
Department of Parasitology, Faculty of Medicine
University of Colombo, Colombo, Sri Lanka

N.R. de Silva
Department of Parasitology, Faculty of Medicine
University of Kelaniya, Ragama, Sri Lanka

I. Hettiarachchi
Plantation, Health & Social Welfare Trust
Colombo, Sri Lanka

Introduction

Sri Lanka is one of the few countries that routinely administers anthelminthics during pregnancy, and it has been doing so for almost 20 years. Very few studies have been done in Sri Lanka to determine the prevalence rates of soil-transmitted nematode infections among adolescent girls and women of child-bearing age (see Table 1). The adverse effects of these infections on growth and nutrition are well known; recently it has been shown that trichuriasis is associated with significantly lower serum vitamin A levels (Atukorala & Lanerolle, 1999), and that even mild infection with *Trichuris* has adverse effects on serum vitamin A concentrations. Studies on iron metabolism suggest that vitamin A deficiency may impair mobilization of iron stores (Bloem et al., 1989). Hookworm infection is known to cause blood loss, sometimes leading to iron-deficiency anaemia. Intervention studies have shown that combined supplements—containing both vitamin A and iron—are more effective than iron alone in eliminating anaemia in women (Suharno & Muhilal, 1996).

Before 1980, treatment of helminthiasis in pregnancy in Sri Lanka was carried out only when stool samples submitted at antenatal clinics were found to be positive for intestinal nematode infections; treatment was not given routinely and coverage was poor. It was only when studies carried out around that time revealed that pregnant women in Sri Lanka had unacceptably high levels (56–78%) of

[1] Professor M.M. Ismail, Department of Parasitology, Faculty of Medicine, University of Colombo, Kynsey Road, Colombo 8, Sri Lanka
Tel: +94 1 695300, Fax: +94 1 699284, E-mail: mismail@slt.lk

Table 1. Prevalence of soil-transmitted nematode infections among adolescent girls and adult women in Sri Lanka

Study	Location	Age group (years)	No. of cases	Prevalence rate (%)		
				Ascaris	Hookworm	*Trichuris*
Sorensen et al. (1996b)	Plantations	18–44	246	69.5	41.4	56.7
Ismail & Naotunne (1997)	Plantations	13–18	161	45.6	29.1	49.2
Atukorala & Lanerolle (1999)	Urban	14–18	383	16.7	5.7	46.2
	Rural	14–18	231	2.2	4.8	4.3

anaemia (Ministry of Health, 1994) that serious attention was paid to the need to control anaemia in pregnancy.

Anthelminthics, anaemia, and pregnancy

In a study in the Colombo area, Atukorala & de Silva (1990) showed that 59% of adolescent females (14–18 years of age) had depleted iron stores and were therefore at risk of developing anaemia during times of higher iron demand such as in pregnancy.

Anaemia predisposes to severe morbidity in pregnant women and reduces tolerance to the normal blood loss during childbirth, which could pose a severe threat to life. Stillbirths and low birth weights are also associated with moderate and severe anaemia during pregnancy. Among the major causes of anaemia, nutritional deficiencies of iron and folate are by far the most common. However, hookworm disease is an important contributory factor in areas where this infection is endemic, especially among women of reproductive age (Pawlowski, Schad & Stott, 1991). In Sri Lanka, iron–folate supplementation had been in place for many years; mebendazole (100 mg twice daily for 3 days) was routinely supplied after the first trimester on an island-wide basis, but compliance was poor. All issues relating to the control of anaemia in pregnancy were re-examined by a Consultative Committee set up by the Ministry of Health and Women's Affairs. Its recommendations (Ministry of Health, 1994) include the use of a single 500-mg dose of mebendazole after the first trimester to control helminthiases in pregnancy. A single dose is much simpler to administer than the 3-day course of mebendazole used previously, and encourages better compliance.

Mebendazole is one of the most widely used anthelminthics in Sri Lanka because of its efficacy, safety, and low cost. The benefits of mebendazole therapy were shown by Atukorala et al. (1994) in a study of pregnant women in a hookworm-endemic area. There was a significant improvement in haemoglobin concentration and iron status when mebendazole was combined with iron–folate supplements compared with iron–folate alone. Mebendazole tablets—in both

Table 2. Cure rates (CR) and egg-reduction rates (ERR) for single-dose anthelminthic drugs

Anthelminthic drug	Sample size	*Ascaris*		Hookworm		*Trichuris*	
		CR (%)	ERR (%)	CR (%)	ERR (%)	CR (%)	ERR (%)
Mebendazole (locally produced)	145	95.8	98.0	28.7	72.0	29.1	31.6
Mebendazole (proprietary)	136	97.6	99.7	35.8	74.5	26.1	61.6
Albendazole (proprietary)	118	97.2	99.6	77.9	95.4	26.2	50.3

100-mg and 500-mg formulations—are now manufactured by the State Pharmaceutical Manufacturing Corporation in Sri Lanka at one-twentieth of the cost of the proprietary drug, which makes it affordable for community treatment programmes. Sorensen et al. (1996a) have shown that the anthelminthic efficacy of locally produced mebendazole is comparable to that of the proprietary preparation (see Table 2).

Although the routine use of mebendazole during pregnancy is now in place, there is still room for improvement. Some pregnant women do not attend antenatal clinics and sometimes the clinics run out of drug supplies. Nevertheless, it is estimated that about 75% of pregnant women receive anthelminthic treatment in Sri Lanka (de Silva et al., 1999).

Safety of anthelminthic therapy

Concerns about the safety of anthelminthic therapy in pregnancy have been expressed from time to time. In 1994, WHO convened an Informal Consultation on Hookworm Infection and Anaemia in Girls and Women to consider this issue. The Consultation promoted the use of anthelminthics in pregnancy after the first trimester, but it also recommended evaluation of the long-term safety—particularly in terms of birth outcomes—of anthelminthic therapy in pregnancy.

In a major cross-sectional retrospective study carried out in Sri Lanka, de Silva et al. (1999) assessed the effect on birth outcomes of mebendazole therapy during pregnancy. The rates of major congenital defects, stillbirth, perinatal death, and very low birth weight (<1500 g) were compared in babies of mothers who had taken mebendazole during pregnancy and babies whose mothers had taken no anthelminthic (controls). The study was carried out in two state-run tertiary care hospitals: all women giving birth at the university obstetric units in the Colombo North Teaching Hospital at Ragama (serving a largely urban community) and the Peradeniya Teaching Hospital (serving a more rural population) were recruited to the study. The primary outcome measure was the frequency of major birth defects, defined as structural or functional defects requiring surgical or medical intervention. Frequencies of stillbirth, perinatal death, and very low birth weight were secondary outcome measures.

Of the 7087 women recruited to the study, 5275 (74.4%) said they had taken a course of mebendazole at least once during the current pregnancy (mebendazole group) while 1737 (24.5%) had taken no anthelminthics (control group). Those who had taken an anthelminthic other than mebendazole and those who could not recall what drug they had taken were excluded from the study. In the mebendazole group, 4890 women (92.7%) had taken a total dose of 600 mg (100 mg twice daily for 3 days). The prescription of mebendazole was verified from antenatal notes in 67% of the mebendazole group (documented exposures). Analyses of certain factors in the two study populations showed that the extent of mebendazole use was similar in Ragama and Peradeniya (74.7% vs 75.8%) as was the baseline (background) frequency of major congenital defects (1.9% vs 1.6%) and of stillbirth and perinatal deaths (2.2% vs 2.3%).

The two study populations showed differences in several factors that are known or suspected to be associated with an increased risk of congenital defects, and data analyses were therefore stratified according to hospital wherever possible. There were 97 (1.8%) babies with major congenital defects in the mebendazole group and 26 (1.5%) in the control group; these frequencies were not statistically different (see Table 3). No significant differences were observed in the Ragama and Peradeniya sub-analyses or when the analysis was limited to documented exposures. The incidence of major congenital defects was slightly higher in women who had taken mebendazole in the first trimester (against current recommendations) than in the control group, but the differences were not significant; this was true overall as well as in Ragama and Peradeniya separately. The overall incidence of major congenital defects in this study (1.75%) is comparable to that reported in other studies (Nevin, 1982; Sirisena, Nanayakkara & Jayasena, 1993).

Rates of stillbirth, perinatal death, and very low birth weight were significantly lower in the mebendazole group than in the controls (Table 4). A high proportion of the stillbirths and perinatal deaths were among babies of very low birth weight, which is to be expected. If antenatal treatment with mebendazole does have a beneficial effect on birth outcome, one possible explanation would be

Table 3. Incidence of major birth defects in mebendazole group and controls

	Mebendazole group	Control group	Odds ratio (95% CI)[a]	*P*
All potential exposures	97/5275 (1.8%)	26/1737 (1.5%)	1.24 (0.8–1.91)	0.39
Documented exposures only	67/3540 (1.9%)	25/1670 (1.5%)	1.31 (0.82–2.09)	0.31

[a] Mantel–Haenszel weighted odds ratio (95% confidence limits).

Table 4. Incidence of stillbirths and perinatal deaths, and low birth weight in mebendazole and control groups

	Mebendazole group	Control group	Odds ratio (95% CI)[a]	*P*
All potential exposures				
Stillbirths and perinatal deaths	99/5275 (1.9%)	58/1737 (3.3%)	0.55 (0.40–0.77)	0.0004
Birth weight <1500 g	59/5271 (1.1%)	40/1735 (2.3%)	0.47 (0.32–0.71)	0.0003
Documented exposures only				
Stillbirths and perinatal deaths	62/3540 (1.8%)	56/1670 (3.4%)	0.52 (0.36–0.75)	0.0005
Birth weight <1500 g	40/3540 (1.1%)	40/1670 (2.4%)	0.43 (0.27–0.67)	0.0002

[a] Mantel–Haenszel weighted odds ratio (95% confidence limits).

through an increase in birth weight. However, increased mebendazole use and the lower proportion of very low birth weight could be linked simply by the fact that both are reflections of better health-seeking behaviour and antenatal care, rather than being causally linked. Data derived from this study are consistent with the long-standing views of medical practitioners in Sri Lanka that mebendazole therapy is safe during pregnancy.

Anthelminthic treatment during adolescence

Although Sri Lanka now has a strategy for combating intestinal helminthiasis during pregnancy by routinely administering mebendazole to all pregnant women after the first trimester, there is room for improvement. The present estimated coverage of about 75% could be increased by ensuring uninterrupted drug supplies and by encouraging greater attendance at antenatal clinics through awareness programmes. Such programmes should be extended to include adolescents, particularly with a view to reducing the prevalence and intensity of intestinal helminthiases among adolescent girls—the future mothers. The plantation sector of Sri Lanka already has a programme of this nature in place.

Nearly one million people live in the plantations in Sri Lanka. Until about a decade ago, their health was poorer than that of the rest of the island's population. Although there has been substantial investment in improving health facilities, housing, water supplies, and sanitation since the mid-1980s, a large proportion of these people still have poor housing and no access to safe water and sanitation. In 1992–1993, a cross-sectional study of 1614 children aged 3–12 years and 246 women aged 18–44 years, randomly selected from 14 state-owned plantations, revealed a wide prevalence of intestinal nematode infections (Sorensen et al., 1996b) (see Table 1). At that time, only pregnant women after the first trimester were given mebendazole routinely.

A major deworming programme began in 1994 in all the plantations (about 400 estates), providing more than 200 000 children up to the age of 12 years with a single 500-mg dose of mebendazole twice a year. The programme, now in its sixth year, has achieved a considerable decline in the intensity and prevalence of intestinal worm infections. To make the programme more effective, biannual deworming was extended in 1997 to include 13–18-year-olds after examination of a small number of stool samples (161 from 10 estates) revealed the following prevalence rates of infection: *Ascaris* 45.6%, hookworm 29.1%, and *Trichuris* 49.2% (Ismail & Naotunne, 1997). This programme should be effective in controlling intestinal helminthiases in adolescent girls in the plantations and in reducing the prevalence of hookworm anaemia. Indeed, although no causal relationship has been established, subsequent studies have shown that the prevalence of anaemia in the plantations, which was 58.2% during pregnancy and 51.9% postpartum (de Silva & Atukorala, 1996), has declined to 25.1% and 35.0% respectively (Atukorala, 1999, unpublished data). Prevalence of anaemia among non-pregnant women was 27.8% (see Table 5). Targeting adolescent girls for routine anthelminthic therapy, especially in areas where soil-transmitted helminthiases are widely prevalent, seems likely to prove to be a beneficial strategy.

Conclusions

The routine administration of mebendazole in pregnancy, after the first trimester, is now well established in Sri Lanka, with an estimated coverage of 75%. With more vigorous efforts to raise health awareness and greater attendance at antenatal clinics, this proportion should increase. The low cost of mebendazole makes the treatment programme sustainable. Studies in Sri Lanka have proved that mebendazole is safe in pregnancy and does not affect birth outcomes in terms of birth defects, stillbirths, perinatal deaths, and very low birth weights. To target women before the onset of pregnancy, it is recommended that adolescent girls be included in community deworming programmes. Such a scheme now exists in the

Table 5. Prevalence of anaemia in the plantations during pregnancy and postpartum

Prevalence of anaemia	1989–1991[a]		1999[b]	
	n	%	*n*	%
Pregnancy (Hb < 110 g/litre)	309	58.2	181	25.1
Postpartum (Hb < 120 g/litre)	108	51.9	100	35.0
Non-pregnant women (Hb < 120 g/litre)	—	—	97	27.8

[a] de Silva & Aturokala, 1996.
[b] Aturokala, 1999 (unpublished data).

plantation sector in Sri Lanka and the feasibility of extending it to the wider community should be explored.

References

Atukorala TMS, de Silva LD (1990) Iron status of adolescent females in three schools in an urban area of Sri Lanka. *Journal of Tropical Paediatrics*, 36:316–321.

Atukorala TMS, Lanerolle P (1999) Soil-transmitted helminthic infection and its effect on nutritional status of adolescent school girls of low socio-economic status in Sri Lanka. *Journal of Tropical Paediatrics*, 45:18–22.

Atukorala TMS et al. (1994) Evaluation of effectiveness of iron-folate supplementation and anthelminthic therapy against anaemia in pregnancy—a study in the plantation sector of Sri Lanka. *American Journal of Clinical Nutrition*, 60:286–292.

Bloem MW et al. (1989) Iron metabolism and vitamin A deficiency in children in Northeast Thailand. *American Journal of Clinical Nutrition*, 50:332–338.

de Silva LDR, Atukorala TMS (1996) Micronutrient status of plantation workers in Sri Lanka during pregnancy and post-partum. *Journal of Obstetric and Gynaecological Research*, 22:239–246.

de Silva NR et al. (1999) Effect of mebendazole therapy during pregnancy on birth outcome. *Lancet*, 353:1145–1149.

Ismail MM, Naotunne TD (1997) *Third evaluation of the bi-annual deworming programme in the plantations.* Colombo, Sri Lanka, Faculty of Medicine.

Ministry of Health (1994) *A national strategy and broad plan of action for the prevention and control of anaemia in pregnancy.* Colombo, Sri Lanka, Ministry of Health & Women's Affairs.

Nevin C (1982) Genetic disorders. *Clinical Obstetrics and Gynaecology*, 9:3–27.

Pawlowski ZS, Schad GA, Stott G (1991) *Hookworm infection and anaemia: approaches to prevention and control.* Geneva, World Health Organization.

Sirisena J, Nanayakkara D, Jayasena L (1993) A study of major birth defects seen at a general hospital. *Kandy Medical Journal*, 2:58–61.

Sorensen E et al. (1996a) The efficacy of three anthelminthic drugs given in a single dose. *Ceylon Medical Journal*, 41:42–45.

Sorensen E et al. (1996b) The prevalence and control of soil-transmitted nematode infections among children and women in the plantations in Sri Lanka. *Ceylon Medical Journal*, 41:37–41.

Suharno D, Muhilal (1996) Vitamin A and nutritional anaemia. *Food and Nutrition Bulletin*, 17:7–10.

Anthelminthic treatment during pregnancy

H. Torlesse,[1] D.W.T. Crompton, L. Savioli
WHO Collaborating Centre for Soil-Transmitted Helminthiases
University of Glasgow, Glasgow, Scotland

Introduction

Millions of pregnant women throughout tropical and non-tropical regions of the world are infected with intestinal nematodes. The most important intestinal nematodes, in terms of their worldwide prevalence and potential for inducing debilitating disease, include hookworms (*Ancylostoma duodenale* and *Necator americanus*), *Trichuris trichiura*, and *Ascaris lumbricoides.* Hookworms alone are carried by an estimated 44 million pregnant women (Bundy, Chan & Savioli, 1995).

Intestinal nematodes inhabit the intestinal tract and interfere with the host's nutritional status by affecting the intake, intestinal absorption, metabolism, and excretion of nutrients (Stephenson, 1987). Pregnancy is a period of elevated nutritional requirements, when the consequences of infection are more likely to be observed. It has been suggested that pregnant women may develop iron-deficiency anaemia, gain insufficient weight, and give birth to underweight infants if, as a result of these infections, their nutritional requirements are not met (Crompton & Whitehead, 1993; Durnin, 1993; Nesheim, 1993). Pregnancy outcomes may also be affected by ectopic migration of adult worms and by vertical transmission of infection to the fetus or suckling infant.

Anthelminthic treatment is regarded as the most effective means of controlling mortality and morbidity due to intestinal nematode infections (WHO, 1995). However, despite WHO recommendations on the use of anthelminthics drugs during pregnancy (WHO, 1996), few endemic countries have included control of intestinal nematode infections in routine antenatal care. The major obstacles to routine anthelminthic treatment in pregnancy have included the concern that the drugs may have teratogenic or genotoxic effects on the fetus, and a lack of information to support the health benefits of treatment on pregnancy outcome.

During the past decade, important issues relating to the use of anthelminthic drugs during pregnancy have been examined in field studies and by the Informal Consultation on Hookworm Infection and Anaemia in Adolescent Girls and Women convened by WHO in 1994 (WHO, 1995). This brief review draws on

[1] Dr H. Torlesse, Helen Keller International, PO Box 6066, Gulshan, Dhaka 1212, Bangladesh
Tel: +880 2 934 8710, Fax: +880 2 811 3310, E-mail: htorlesse@hotmail.com

the available evidence to examine the importance of intestinal nematode control in antenatal care in areas endemic for these infections.

Intestinal nematode infection and pregnancy outcomes

Maternal and fetal nutrition

Anaemia is a major threat to safe motherhood, and studies linking intestinal nematode infections with iron deficiency and anaemia provide the major argument for anthelminthic treatment during pregnancy. Infants of mothers who are anaemic during pregnancy are more likely to experience adverse outcomes including low birth weight, preterm delivery, and perinatal mortality (Allen, 1997). Small infants have lower iron stores and may develop iron deficiency and anaemia in infancy or early childhood.

Intestinal nematode infections contribute to anaemia by causing intestinal blood loss and by affecting the supply of nutrients for the production of red blood cells. Hookworms are the leading cause of abnormal blood loss in the tropics and subtropics (Fleming, 1989) and, in areas where they are endemic, are likely to be an important cause of iron deficiency and anaemia among women of childbearing age (WHO, 1996). *Trichuris trichiura* also causes intestinal blood loss, although much less so than hookworms on a per-worm basis; however, anaemia may result from gross haemorrhage due to dysentery or rectal prolapse (Bundy & Cooper, 1989). *Ascaris lumbricoides* interferes with the utilization of vitamin A, which is required for haematopoiesis, and all three intestinal nematodes may reduce the intake and absorption of iron and other haematopoietic nutrients by causing anorexia, vomiting, and diarrhoea.

The intensity of hookworm infection, as estimated by faecal egg counts, was found to be negatively associated with haemoglobin concentration in pregnant women in Liberia (Jackson & Jackson, 1987) and Sierra Leone (Torlesse, 1999). Hookworm intensity was also reported to be a dominant risk factor for anaemia in multigravidae in coastal Kenya (Shulman et al., 1996). In Sri Lanka, where only 8.6% of pregnant women were infected with hookworm, there was no correlation between intensity and haemoglobin concentration (WHO, 1995). The latter finding may reflect the low prevalence of hookworm infection among Sri Lankan pregnant women, as there is unlikely to be a relationship between infection and indicators of iron status where the prevalence of infection is below 50% (Holland, 1987).

In Guatemala, pregnant women with intestinal nematode infections were found to gain significantly less weight during pregnancy than uninfected women (Villar, Klebanoff & Kestler, 1989). Furthermore, chronically malnourished pregnant women of short stature with one or more infections had high rates of intrauterine growth retardation. Up to 10% of intrauterine growth retardation among chronically malnourished women was attributable to infection.

Findings elsewhere indicate that intestinal nematode infections do not pose a threat to maternal body size or composition. Three studies in south-east Asian pregnant women found no association between positive infection status and weight gain during pregnancy (D'Alauro et al., 1985; Roberts et al., 1985; Saowakontha et al., 1992). In Sierra Leone, first trimester measurements of height, weight, body mass index, mid-upper arm circumference, and skinfolds were not associated with the prevalence or intensity of intestinal nematode infections, and intervention with albendazole did not appear to have any impact on the change in anthropometric values between the first and third trimesters (Torlesse, 1999). The absence of a relationship between infection and nutritional status in these studies may be explained by the fact that infections were mainly of low intensity. Alternatively, the impact of infection on anthropometric measures of nutritional status may have been concealed by physiological or behavioural adaptation to infection, such as reduced physical activity (Durnin, 1993).

Vertical transmission

Nematode species with obligatory migratory tissue stages, including *A. lumbricoides* and hookworms, have the potential for vertical transmission from the mother to the fetus or suckling infant. There is direct evidence that non-human mammalian hosts acquire related species by transplacental (Loke, 1982) and transmammary routes (Miller, 1981; Burke & Robertson, 1985), but the evidence in humans hosts is circumstantial. Congenital infection with *A. lumbricoides* (Chu et al., 1972; da Costa-Macedo & Ray, 1990) and *A. duodenale* (Nwosu, 1981; Sen-Hai & Wei-Xia, 1990) is indicated by the recovery of adult worms or eggs from infants younger than the pre-patent period of the nematodes. There is also evidence of a transmammary route of infantile *A. duodenale* infection (Schad, 1990, 1991), but larvae have yet to be identified in human breast milk. Vertical transmission of *N. americanus* is not expected to occur since this nematode does not enter a period of parental hypobiosis, a prerequisite for vertical transmission (Schad, 1991).

While vertical transmission appears to account for a very small proportion of the total case-load, the implications of infection for the very young are more likely to be severe. In view of their small body size, infants are at greater risk of intestinal obstruction due to infestation with adult *A. lumbricoides* worms and of fatal haemorrhage due to severe *A. duodenale* infection. In China, neonatal and infantile hookworm infection is associated with diarrhoea, anorexia, listlessness, oedema, anaemia, and, in severe cases, death due to intestinal haemorrhage or cardiac failure (Sen-Hai & Wei-Xia, 1990).

Ectopic migration

Ectopic migration of adult *A. lumbricoides* worms within the human host can result in life-threatening complications. Worms may migrate to many organs, including the appendix, bile duct, pancreas, and female genital tract. Invasion of these organs may result in appendicitis, cholangitis, liver abscesses, and pancreatitis (Pinus, 1985; MacLeod, 1988). Although these complications are most frequently observed in children, ectopic migration may be induced in pregnant women by the stress of labour or inhalation anaesthetics (MacLeod, 1988). Migration of a single adult worm from the intestine into the trachea and bronchi was reported as the cause of asphyxiation and death in a pregnant woman under anaesthesia during a caesarean section (MacLeod & Lee, 1988).

Safety of anthelminthic treatment

The WHO Model List of Essential Medicines (WHO, 2003) includes four anthelminthic drugs: albendazole, mebendazole, levamisole, and pyrantel. All four drugs are known to be highly efficacious and to have minimal side-effects but data about their use in pregnancy are extremely limited. Clinicians and public health workers are faced with a difficult challenge: should anthelminthic drugs be given to pregnant women known to be infected with hookworm? There is no evidence to indicate that either levamisole or pyrantel is teratogenic or genotoxic, but both albendazole and mebendazole have been shown experimentally to have teratogenic effects in some species of laboratory animals. However, these effects have not been observed at clinically relevant doses in farm animals. Species differences in pharmacokinetics and metabolism may protect against teratogenic or genotoxic effects in humans.

After a thorough review of information on the properties of the four anthelminthic drugs and with regard to the seriousness of hookworm infection during pregnancy, WHO's Informal Consultation concluded that the risks of delaying anthelminthic treatment outweigh the teratogenic risks to the fetus in areas where hookworm infection is endemic (prevalence greater than 20–30%) and anaemia is prevalent (WHO, 1996). In these areas, WHO supports treatment with any one of the four anthelminthic drugs after the first trimester of pregnancy. Recent guidelines on the prevention and treatment of iron-deficiency anaemia during pregnancy (Stoltzfus & Dreyfus, 1998) reinforce WHO's recommendation. If hookworms are highly endemic (prevalence >50%), anthelminthic treatment should be repeated in the third trimester of pregnancy (Stoltzfus & Dreyfus, 1998). Anthelminthic treatment at or immediately after delivery is probably the most effective approach to reducing the vertical transmission of *A. duodenale* (WHO, 1996).

Following the WHO Informal Consultation, an important study in Sri Lanka, where anthelminthic treatment is routinely given to pregnant women after the first trimester, improved our knowledge of the consequences of using anthelminthic drugs during pregnancy. The study involved a prospective, unmatched, case–control investigation, which compared the birth defect rates in children born to 5725 mothers who had taken mebendazole during pregnancy with the rates in children born to 1737 mothers who had not taken any anthelminthic drug (de Silva et al., 1999). There was no significant difference in the birth defect rates among children from the two sets of mothers. Contrary to medical advice given at the time, 407 women had knowingly taken mebendazole during the first trimester, but again the birth defect rate for the children of these mothers was not significantly different from that of children born to untreated mothers.

A second study in Sierra Leone followed the birth outcomes of 125 women, 61 of whom had received albendazole in the second trimester of pregnancy (Torlesse, 1999; Torlesse & Hodges, 2001). No evidence was found to indicate that the anthelminthic drug had affected birth defect rates.

Effects of anthelminthic treatment

The effects of intestinal nematode infection on maternal and fetal health suggest that anthelminthic treatment during pregnancy may have several beneficial effects, including reductions in maternal and infant anaemia, low birth weight, and maternal and perinatal mortality (see Table 1). Of all these potential outcomes, only maternal anaemia has been studied in detail.

A study in Sri Lanka examined the impact of mebendazole treatment during pregnancy on maternal iron status and haemoglobin concentration (Atukorala et al., 1994). A single course of mebendazole treatment, in addition to iron–folate supplements, resulted in a significant improvement in iron stores and haemoglobin concentration compared with iron–folate supplements alone. The prevalence and intensity of intestinal nematode infections were not measured in these women. A separate parasitological investigation among 146 non-pregnant women

Table 1. Potential benefits of anthelminthic treatment for pregnant women and their infants

	Potential benefit
Pregnant women	Increased maternal iron stores and haemoglobin concentrations Reduced need for blood transfusion Reduced maternal mortality associated with severe anaemia Improved maternal weight gain Prevention of ectopic migration of *Ascaris lumbricoides*
Infants	Improved fetal growth Prevention of preterm delivery Reduced low birth weight Improved iron stores and total haemoglobin mass Prevention of vertical transmission of *Ancylostoma duodenale* Reduced perinatal mortality Improved early childhood development

in Sri Lanka found evidence of hookworm infections in 41.4%, the majority of which were mild infections with mean faecal egg counts <1000 (Atukorala et al., 1994).

The efficacy of albendazole and iron–folate supplements in controlling maternal iron deficiency and anaemia was investigated in western Sierra Leone with a view to determining priorities for intervention (Torlesse, 1999; Torlesse & Hodges, 2001). A cohort of 125 pregnant women were randomly assigned to receive albendazole or the control (calcium) in the second trimester, plus iron–folate supplements or the control (vitamin D) in the second and third trimesters. At baseline, in the first trimester of pregnancy, the prevalence of intestinal nematode infections was as follows: *N. americanus* 65.6%, *T. trichiura* 74.4%, and *A. lumbricoides* 20.0%. The majority of women harboured light infections: only 10.3% had faecal hookworm egg counts >1000. Anaemia was diagnosed in 56.0% of the women and was associated with iron deficiency (serum ferritin <20 μg/litre) in 21.2%.

Albendazole was highly effective in reducing the prevalence and intensity of *A. lumbricoides* and *N. americanus*; it was ineffective in clearing *T. trichiura* infections but reduced the geometric mean egg counts considerably. The protection afforded by a single dose of albendazole, in terms of reducing the intensity of infection below the level associated with morbidity, extended for the duration of pregnancy for all infections. The interventions were not accompanied by improvements in haemoglobin or serum ferritin concentrations during pregnancy, but these values did not worsen in women given both albendazole and the supplements. After controlling for baseline haemoglobin concentration and the season, the mean decline in haemoglobin concentration between the first and third trimester in women given albendazole was 6.6 g/litre less than in women given the albendazole control; the corresponding value for iron–folate supplements was 13.7 g/litre haemoglobin.

The findings from Sri Lanka and Sierra Leone indicate that anthelminthic treatment, by reducing intestinal iron loss and lowering iron requirements, complements iron–folate supplementation. The implication of these findings for pregnant women in other regions will depend on the local prevalence and intensity of intestinal nematode infections, the extent of iron deficiency and anaemia, and the presence of other factors that contribute to anaemia.

Control of intestinal nematodes is included in national strategies for the prevention and control of maternal anaemia during pregnancy in the Seychelles, Sierra Leone, and Sri Lanka. Besides the impact on iron status and haemoglobin levels, few of the potential effects of anthelminthic treatment have been adequately studied, which may explain the failure of many countries to give priority to intestinal nematode control during pregnancy. More information on the health benefits of anthelminthics, particularly on infant outcomes, is needed to build up more persuasive arguments for their use during pregnancy.

Conclusions

Intestinal nematode infections and undernutrition frequently coincide during pregnancy in developing countries. These infections, in particular hookworm infections, contribute to iron deficiency and anaemia during pregnancy. Anthelminthic treatment has been shown to be effective in improving maternal iron stores and haemoglobin levels. There is also some evidence to suggest that infection may contribute to subnormal fetal growth, although this relationship has not been confirmed by intervention studies. Previous concerns about the teratogenic or genotoxic effects of anthelminthic drugs are no longer valid, but as a general rule, treatment should be administered after the first trimester of pregnancy. More countries endemic for hookworm infection should be encouraged to include intestinal nematode control in routine antenatal care, as anthelminthic treatment reduces the health risks attached to low haemoglobin levels and could potentially reduce the incidence of low birth weight and its sequelae. Arguments for intestinal nematode control during pregnancy will be greatly strengthened if the merits of anthelminthic treatment for the fetus as well as the mother are more clearly defined.

References

Allen LH (1997) Pregnancy and iron deficiency: unresolved issues. *Nutrition Reviews*, 55:91–101.

Atukorala TMS et al. (1994) Evaluation of effectiveness of iron-folate supplementation and anthelminthic therapy against anaemia in pregnancy—a study in the plantation sector of Sri Lanka. *American Journal of Clinical Nutrition*, 60:286–292.

Bundy DAP, Chan MS, Savioli L (1995) Hookworm infection in pregnancy. *Transactions of the Royal Society of Tropical Medicine and Hygiene*, 89:521–522.

Bundy DAP, Cooper ES (1989) *Trichuris* and trichuriasis in humans. *Advances in Parasitology*, 28:108–173.

Burke TM, Robertson EL (1985) Prenatal and lactational transmission of *Toxocara canis* and *Ancyclostoma caninum*: experimental infection of the bitch at mid-pregnancy and at parturition. *International Journal for Parasitology*, 15:485–490.

Chu W-G et al. (1972) Neonatal ascariasis. *Journal of Pediatrics*, 87:783–785.

Crompton DWT, Whitehead RR (1993) Hookworm infections and human iron metabolism. *Parasitology*, 107(Suppl.):S137–S145.

da Costa-Macedo LM, Ray L (1990) *Ascaris lumbricoides* in neonates: evidence of congenital transmission of intestinal nematodes. *Revista da Instituto de Medicina Tropical de São Paulo*, 32:351–354.

D'Alauro F et al. (1985) Intestinal parasites and pregnancy. *American Journal of Obstetrics and Gynecology*, 66:639–643.

de Silva et al. (1999) Effect of mebendazole therapy during pregnancy on birth outcomes. *Lancet*, 353:1145–1149.

Durnin JVGA (1993) Energy requirements in human pregnancy and parasitic infection. *Parasitology*, 107(Suppl.):S169–S175.

Fleming AF (1989) Tropical obstetrics and gynaecology. 1. Anaemia in pregnancy in tropical Africa. *Transactions of the Royal Society of Tropical Medicine and Hygiene*, 83:441–448.

Holland C (1987) Hookworm infection. In: Stephenson LS, Holland C, eds. *Impact of helminth infections on human nutrition. Schistosomes and intestinal nematodes.* London, Taylor & Francis:128–160.

Jackson RT, Jackson LC (1987) Biological and behavioural contributors to anaemia during pregnancy in Liberia, West Africa. *Human Biology*, 59:585–597.

Loke YW (1982) Transmission of parasites across the placenta. *Advances in Parasitology*, 21:155–228.

MacLeod CL (1988) Intestinal nematode infections. In: MacLeod CL, ed. *Parasitic infections in pregnancy and the newborn*. Oxford, Oxford University Press:192–215.

MacLeod CL, Lee RV (1988) Parasitic infections. In: Burrow GN, Ferris TF, eds. *Medical complications during pregnancy*. Philadelphia, W.B. Saunders:425–447.

Miller GC (1981) Helminths and the transmammary route of infection. *Parasitology*, 82:335–342.

Nesheim MC (1993) Human nutrition needs and parasitic infection. *Parasitology*, 103(Suppl.): S7–S18.

Nwosu ABC (1981) Human neonatal infections with hookworms in an endemic area of Southern Nigeria—a possible transmammary route. *Tropical and Geographical Medicine*, 33:105–111.

Pinus J (1985) Surgical complications of ascariasis in Brazil. In: Crompton DWT, Nesheim MC, Pawlowski ZS, eds. *Ascariasis and its public health significance*. London, Taylor & Francis:161–165.

Roberts NS et al. (1985) Intestinal parasites and other infections during pregnancy in southeast Asian refugees. *Journal of Reproductive Medicine*, 30:720–725.

Saowakontha S et al. (1992) Nutritional health and parasitic infection of rural Thai women of child-bearing age. *Nutrition Research*, 12:929–942.

Schad GA (1990) Hypobiosis and related phenomena in hookworm infection. In: Schad GA, Warren KS, eds. *Hookworm disease: current status and new directions*. London, Taylor & Francis:71–88.

Schad GA (1991) Hooked on hookworm: 25 years of attachment. *Journal of Parasitology*, 77:179–186.

Sen-Hai Y, Wei-Xia S (1990) Hookworm infection and disease in China. In: Schad GA, Warren KS, eds. *Hookworm disease: current status and new directions*. London, Taylor & Francis:44–54.

Shulman CE et al. (1996) Malaria is an important cause of anaemia in primigravidae: evidence from a district hospital in coastal Kenya. *Transactions of the Royal Society of Tropical Medicine and Hygiene*, 90:535–539.

Stephenson LS (1987) Human malnutrition and helminths. In: Stephenson LS, Holland C, eds. *Impact of helminth infections on human nutrition. Schistosomes and intestinal nematodes*. London, Taylor & Francis:1–20.

Stoltzfus RJ, Dreyfus ML (1998) *Guidelines for the use of iron supplements to prevent and treat iron deficiency anaemia*. Washington, DC, International Life Sciences Institute.

Torlesse H (1999) *Parasitic infection and anaemia during pregnancy in Sierra Leone* [Thesis]. Glasgow, University of Glasgow.

Torlesse H, Hodges M (2001) Anthelminthic therapy and reduced decline in haemoglobin concentrations during pregnancy (Sierra Leone). *Transactions of the Royal Society of Tropical Medicine & Hygiene*, 95:195–201.

Villar J, Klebanoff M, Kestler E (1989) The effect on fetal growth of protozoan and helminthic infection during pregnancy. *Obstetrics and Gynecology*, 74:915–920.

WHO (1995) *Report of the WHO Informal Consultation on Hookworm Infection and Anaemia in Girls and Women, Geneva, 5–7 December 1994*. Geneva, World Health Organization (document WHO/CTD/SIP/96.1).

WHO (1996) *Report of the WHO Informal Consultation on the Use of Chemotherapy for the Control of Morbidity due to Soil-Transmitted Nematodes in Humans, Geneva, 29 April to 1 May 1996*. Geneva, World Health Organization (document WHO/CTD/SIP/96.2).

WHO (2003) *Selection and use of essential medicines. Report of the WHO Expert Committee (including the 12th Model List of Essential Medicines)*. Geneva, World Health Organization (WHO Technical Report Series, No. 914).

The efficacy of anthelminthics: past, present, and future

J. Horton[1]
GlaxoSmithKline, Brentford, England

Introduction: from empiricism to modern medicines

Helminths have been a problem since time immemorial, and now infect at least a quarter of the world's population (Stoll, 1947; Bundy, 1997). They are perhaps the first disease-causing organisms to have been demonstrated in humans: eggs—probably of *Enterobius vermicularis*—have been found in burial sites dating back 10 000 years. Moreover, since helminths, especially *Ascaris lumbricoides*, are visible to the naked eye, they are referred to in some of the earliest medical writings. The Ebers Papyrus, a Pharaonic medical codex of some 3500 years ago, records a number of different worms infecting humans, including hookworm and schistosomes, together with treatments to get rid of them (Chandler, 1929). While these treatments would seem bizarre today, some of the ingredients may have been effective. Later medical literature from Greece, Rome, Sumeria, and India also describes treatments that may well have been effective, being based on extracts of the leaves, bark, and roots of particular plants, together with various minerals; however, some of these mixtures were probably quite toxic—to both worms and patients. Similar knowledge developed (or was perhaps regained, especially from Roman and Arabic texts) during the Middle Ages in Europe, when the first "scientific" formularies were drawn up. Among the treatments recommended was wormwood (*Artemesia absinthum*) which, in addition to its more notorious use as a hallucinogen, was used to expel worms, and *Dryopteris felix-mas* which was used for tapeworm (Mabey, 1996). These were usually combined with strong purgatives and sometimes with more toxic substances, such as mercury salts (calomel).

The first purely synthetic anthelminthics were developed in the early years of the 20th century, but their use was based largely on trial and error or chance observation. In general they were simple molecules, such as tetrachloroethylene or hexylresorcinol, and frequently quite toxic, especially as they had to be used for several days to have an effect. The first mass treatment programme was established by the Rockefeller Foundation to eliminate hookworm disease in the southern United States, and made extensive use of resorcinol. Starting immediately

[1] Dr J. Horton, TDR/PRD, World Health Organization, Geneva, Switzerland (present address)
Tel: +41 22 791 2698, E-mail: hedgepigs@aol.com

before the First World War, the programme was reportedly successful, and it is the Rockefeller reports that provide early evidence of both the deleterious effects of hookworm infection and the benefits of treatment (Rockefeller Foundation, 1922).

In the subsequent quarter century there were few advances, and it was not until 1947, after the Second World War, that the first safe and effective anthelminthic—piperazine—was developed. This drug proved particularly useful for the treatment of *Enterobius* infections, and is still used today for that purpose. It was also quite effective against *Ascaris*, but had little effect on other intestinal nematodes. Pyrivinium (1958) was also developed to treat *Enterobius* infection but had little effect on other helminths, while bephenium salts (1959) were active principally against hookworm and were used for a while in mass treatment campaigns.

In 1962, tiabendazole, a benzimidazole, was introduced as the first of the broad-spectrum anthelminthics. Unlike later compounds, it is a hydrochloride and is relatively well absorbed, but it is also fairly toxic at therapeutic doses, causing systemic effects such as nausea. It proved very effective against a range of helminths, including *Ascaris*, *Enterobius*, hookworm, and particularly *Strongyloides stercoralis*, for which it remained the treatment of choice until the development of ivermectin. It is relatively ineffective against *Trichuris tichiura*, but has good efficacy against systemic worm infections, especially strongyloidiasis. As a topically applied suspension, tiabendazole has also been widely used for the treatment of cutaneous larva migrans. Today, tiabendazole has been largely replaced by more recent anthelminthics, all of which are still widely used.

The search for effective medicines has led to safer and easier to use compounds. Pyrantel (1968) and levamisole (1969) were the first well-tolerated anthelminthics, generally used in single, weight-based doses. Pyrantel is effective against hookworms (particularly *Ancylostoma duodenale*) and, when used in combination with oxantel (1972), against *Trichuris*, and has remained one of the mainstays of therapy. Levamisole on the other hand, though well tolerated, has fallen out of use except in areas where *Ascaris* is the principal helminth species.

The most recent developments have been the result of a major thrust in the veterinary anthelminthic field which produced several products derived from tiabendazole, including the broad-spectrum benzimidazole carbamates, mebendazole (1973), its *p*-fluoro derivative flubendazole (1976), and albendazole (1981). Both mebendazole and flubendazole were originally used in divided doses over 2–3 days, but mebendazole is now largely used as a single dose for common intestinal infections. Albendazole has always been used as a single-dose anthelminthic. A number of other benzimidazoles have been produced for use in animals, but only limited data are available on their efficacy and safety in human helminth infections. Thus, despite nearly half a century of drug development, only three compounds (pyrantel, mebendazole, albendazole) are now in regular use, with levamisole and flubendazole being used in a more limited manner.

How good are the current products?

There are considerable problems in evaluating the efficacy of many of the older compounds, as the methods used often varied considerably and are poorly described in the available reports. Janssens (1986) provided a very detailed compilation of efficacy data for many of the older compounds, and a considerable amount of early data on the modern compounds as well. In the past 20 years, however, standardized protocols produced by the World Health Organization, combined with well-established diagnostic methods such as the Kato–Katz technique, have made comparison easier. Moreover, there has been a tendency to undertake larger studies to rule out the statistical sampling errors previously encountered. Surprisingly, there are relatively few comparative clinical studies of modern compounds.

Perhaps the best recent study, which complies with the most exacting standards of clinical evaluation, is a large comparison of albendazole and mebendazole conducted in Pemba Island, United Republic of Tanzania (Albonico et al., 1994). The results (see Table 1) show excellent efficacy for both compounds against *Ascaris* in terms of cure and egg reduction rates. In contrast, albendazole was the more effective against *Necator americanus*; neither compound was particularly effective against *Trichuris*, although egg reduction rates were quite high. Even with relatively modest cure rates, morbidity may decline if worm burdens are reduced sufficiently, and—since benzimidazoles have effects on the eggs and larvae of helminths—transmission may also be reduced.

Bennett & Guyatt (2000) recently evaluated a number of efficacy studies of albendazole (single 400-mg dose) and mebendazole (both single 500-mg and 3-day dosing) against the common helminths (see Table 2). Their conclusions, based on data from different parts of the world, largely confirm the data from the Pemba Island study. They confirm the high efficacy of both compounds against *Ascaris* and show that single-dose treatment with albendazole is more effective against hookworm; albendazole and mebendazole have similar, but poor, efficacy against

Table 1. Comparison of the efficacy of single-dose albendazole and mebendazole treatment of children with intestinal helminthiasis: Pemba Island, United Republic of Tanzania[a]

Helminth	Treatment	Cure rate (%)	Egg reduction rate (%) (95% CI)
Ascaris	Albendazole	98.9	99.6 (99.4–99.7)
	Mebendazole	97.8	99.3 (99.2–99.5)
Hookworm	Albendazole	56.8	97.7 (97.1–98.1)
	Mebendazole	22.4	82.4 (78.8–85.5)
Trichuris	Albendazole	10.5	73.3 (69.4–76.8)
	Mebendazole	14.2	81.6 (78.8–84.1)

[a] Source: Albonico et al. (1994); 1174 children treated with albendazole, 1120 children treated with mebendazole.

Table 2. Comparison of the anthelminthic efficacy of albendazole and mebendazole[a]

Helminth	Drug[b]	Cure rate (%)			Egg reduction rate (%)		
		Median	Upper quartile	Lower quartile	Median	Upper quartile	Lower quartile
Trichuris	A	35	68	22	80	87	43
	M	38	74	25	82	87	72
	Mx3	75	85	66	93	97	73
Ascaris	A	97	100	89	99+	100	98
	M	97	100	92	99	100	98
	Mx3	98	100	95	99+	100	98
Hookworm	A	85	85	70	87	92	75
	M	30	73	20	84	100	55
	Mx3	80	90	74	94	99	85

[a] Source: Bennett & Guyatt (2000).
[b] A = albendazole, 400 mg; M = mebendazole, 500-mg single dose; Mx3 = mebendazole, 100 mg twice daily for 3 days.

Trichuris. Interestingly, the efficacy of 3-day mebendazole treatment is significantly better than that of single-dose treatment in hookworm infection, approximating to that of single-dose albendazole treatment. The analysis does not distinguish between the two hookworm species, since there is known to be a marked difference in the response of *Ancylostoma* and *Necator* to mebendazole, which is not the case with albendazole (Holzer & Frey, 1987). The efficacy of mebendazole against *Trichuris* is also increased substantially with a 3-day regimen. Despite the good results, the 3-day regimen is not practical for field use or mass treatment campaigns, although it would be useful for treatment of individuals. One other interesting observation in this study was an apparent geographical variation in albendazole efficacy against *Trichuris*, which was greater in Africa and America than in Asia.

The Bennett & Guyatt analysis used study material largely from single compound investigations, and comparison is therefore more open to methodological variation between studies and to geographical differences in infection intensity and prevalence of the different helminth species. The following analysis, summarized in Table 3, used data from comparative studies only, and variations caused by differences in population size were corrected using a weighting process (Der Simonian & Laird, 1986). Although the number of studies available for analysis was much smaller, the findings were very similar to those of Bennett & Guyatt, showing albendazole to be more effective in hookworm infections and 3-day mebendazole treatment to be better for *Trichuris* infections than single-dose treatment. It appears that the efficacy of benzimidazoles is variable against hookworm (especially *Necator*) and generally poor against *Trichuris*, particularly with single doses.

Table 3. Comparison of the efficacy of albendazole and mebendazole, using data from comparative studies and weighted estimate to correct for sample-size differences[a]

Helminth	No. of studies	Mebendazole dose	Weighted estimate[b] (albendazole – mebendazole)	95% confidence Intervals
Necator	2	400–600 mg, single dose	0.45	0.23, 0.67
All hookworms	6	400–600 mg, single dose	0.47	0.35, 0.59
Ascaris	6	400–600 mg, single dose	0.007	–0.039, +0.053
Trichuris	8	400–600 mg, single dose	–0.097	–0.23, +0.04
Trichuris	2	100 mg, twice daily for 3 days	–0.27	–0.43, –0.11

[a] Data from Ioli et al. (1982), Amato-Neto, Moreira & Campos (1983), Cruz (1983), Wang (1987), Sinniah, Chew & Sobramaniam (1990), Bartoloni et al. (1993), Jonksuksuntigul et al. (1993), Albonico et al. (1994).
[b] Weighted estimates by method of Der Simonian & Laird (1986); positive values show albendazole to be superior to mebendazole.

There are relatively few recent data on either pyrantel or levamisole. Both drugs are clearly effective against *Enterobius* and *Ascaris*, with cure rates above 90% reported almost universally. They are less effective against hookworm: cure rates in excess of 80% are seen in *Ancylostoma* infections, but for *Necator* efficacy falls to around 60% or even lower (Pugh, Teesdale & Burnham 1989). The most recent data on pyrantel efficacy against *Necator* are very similar to those for mebendazole (Sacko et al., 1999). In *Trichuris* infections, efficacy is poor, with cure rates usually below 50%.

All currently used anthelminthics give poor results with *Trichuris*, and it is therefore of interest that oxantel, a pyrantel analogue, historically showed consistently good results. Oxantel is poorly absorbed from the gastrointestinal tract (Garcia, 1976) and high concentrations are therefore achieved in the lower bowel. Single doses produce cure rates of over 80% and egg reductions in excess of 90%. However, the drug has little effect on helminths inhabiting the upper gastrointestinal tract, especially *Ascaris*, and is therefore used in combination with pyrantel (Sinniah, Chew & Sobramaniam, 1990). Unfortunately, there are few recent data on this combination, which warrants further investigation using modern techniques.

Generics—are they a good option?

Anthelminthics are of sufficient age that they are no longer covered by patents and so can be freely copied and produced as generic formulations. A generic version is often cheaper than the original proprietary product, which is a clear incentive for its purchase and use, either for individual treatment or—more significantly—for community treatment programmes. Most data on efficacy and safety data derive from research on the original proprietary products and were the basis for licensing these drugs for use in humans. It is therefore crucial that each generic product not only contains the correct amount of active ingredient but also behaves in a similar manner to the original drug product.

Many anthelminthics are notoriously difficult to formulate correctly, since they are poorly water-soluble. An examination of generic preparations of albendazole from around the world has highlighted this problem. Most products examined contained the stated amount of compound. However, only about 20% had pharmaceutical characteristics similar to those of the originator drug (Galia, Horton & Dressman, 1999). Following general pharmacopoeial rules, the two most commonly used tests are of disintegration (the time taken for tablets to fragment under standard conditions) and dissolution (the time taken for a given amount of the active compound to go into solution).

Disintegration tests are the simplest to conduct, and limits for the time to disintegration are well established—less than 15 minutes for uncoated tablets. In an analysis of albendazole generics, about one in five products failed these tests, with disintegration times exceeding 30 minutes (Gallia, Horton & Dressman, 1999; Horton, personal communication). The worst examples showed little physical change after 2 or 3 hours, and could therefore not be expected to be effective against helminths found principally in the upper intestine, such as *Ascaris* and hookworm. These tablets also failed dissolution tests, with little drug released over the test period (<85% of drug in solution after 30 minutes, <90% after 45 minutes). More surprising is the fact that about half the tablets that disintegrated within the standard time failed on the dissolution test (see Figure 1). There are no simple explanations for this: there may be different reasons for each product. With low-solubility compounds, the wetting of material is important and depends on the excipients used. Other factors, such as original particle size (larger particles dissolve less easily) or over-compression of tablets (causing a change in crystal structure), may also play a part.

In assessing any generic product it is therefore important, irrespective of the cost of the product, to establish whether it is behaving in a pharmaceutically acceptable manner. For poorly soluble compounds, the disintegration test (British Pharmacopoeia) by itself is not useful, and the dissolution test (United States Pharmacopeia) is to be preferred. The amount of active compound should be within certain limits (say, ±10% of quoted amount) and should dissolve in a reasonable time (say, >80% in 45 minutes). A compound falling within these limits could be expected to behave in a similar manner to the originator compound.

Suspensions might be expected to perform better than tablets, but a review of generics has revealed different problems. These include difficulties in reconstitut-

Figure 1. Behaviour of 20 albendazole generics in dissolution and disintegration tests

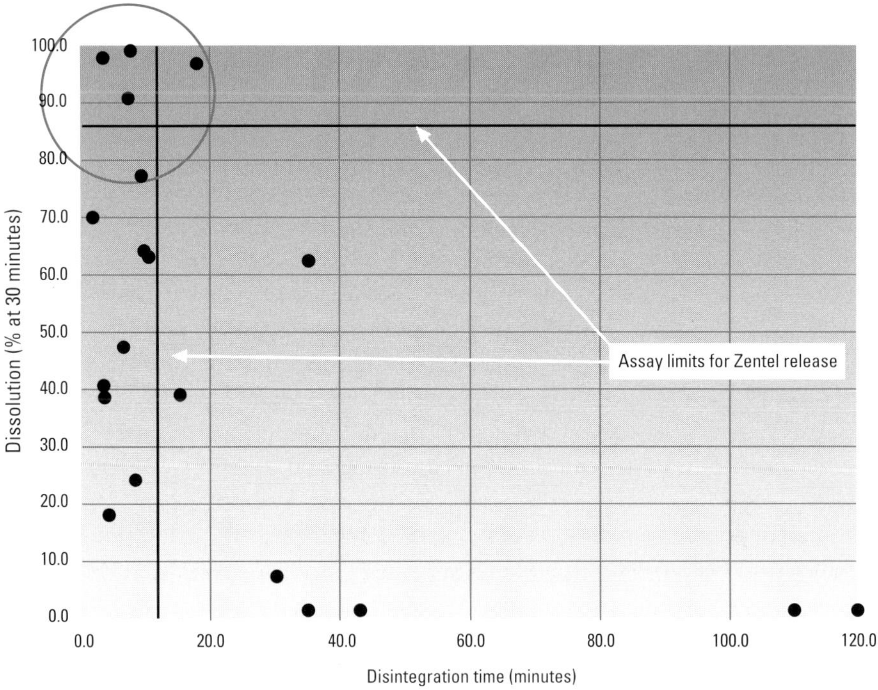

The approved release characteristics are indicated. Only four of the generics fall within the acceptable limits. A doubling of these limits would include only one further product. Zentel is albendazole produced by GlaxoSmithKline.

WHO 03.66

ing the suspension (on standing, the material packs into a solid mass that cannot be resuspended), problems with stability (pH is critical in this respect), and inadequate quantities of active ingredient (almost 50% did not contain the stated amount).

A generic product may be preferable in terms of cost, provided that it does the job for which it is intended and that its performance is reasonably comparable to that of the originator compound. One of the potential problems resulting from failure to check the quality and activity of a product is that a large expenditure (such as that entailed in community treatment programmes) will be wasted if treatment fails. In addition, there is the danger that under-treatment will encourage the development of drug resistance.

Drug resistance—threat or reality?

The fear of drug resistance developing in the soil-transmitted helminths that infect humans is based on the experience of veterinary medicine. Benzimidazoles—and, more recently, ivermectin—have been extensively used to eliminate nematodes in cattle, sheep, and other domestic animals, which are healthier and gain useful weight more rapidly after treatment. Treatment is frequent in animal husbandry, often every 6–8 weeks, and resistant strains are selected. Among humans, only certain groups are treated regularly, and the concentration of eggs and larvae to which they are exposed in the environment is lower; thus drug pressure is lower. Furthermore, because treatment is not universal, there will always be substantial numbers of untreated helminths to repopulate the gene pool. Even so, there is some evidence that resistance may occur in human helminth populations in certain circumstances.

Of two studies that report potential drug resistance, the first is an investigation of pyrantel and albendazole conducted in Western Australia (Reynoldson et al., 1997, 1998). Pyrantel had been used for many years to treat Aboriginal populations for hookworm (*Ancylostoma*), and efficacy was initially close to 100%. Treatment was undertaken in closed communities and thus reinfection rates were high. Investigation some 10 years after the start of the programme showed that the efficacy of pyrantel had fallen (15% cure rate and 45% egg-reduction rate), while that of albendazole—which had not been used before—was 100%. The recommended treatment was therefore changed, but not the conditions of use: albendazole is being used four times a year and the potential for development of resistance to a benzimidazole remains.

The second study (Sacko et al., 1999) is more difficult to interpret. It was conducted in an area of Mali where mebendazole was widely available and widely used. In an initial study (de Clercq et al., 1997), mebendazole showed poor results against hookworm (*Necator*). A second study was therefore undertaken, in which both placebo and albendazole were included. The results showed that albendazole had the greatest efficacy in terms of both cure rates and reduction in egg counts, followed—in descending order—by mebendazole, pyrantel, and placebo (see Table 4). Generally speaking, these results are much as might be expected, since neither mebendazole or pyrantel is particularly effective against *Necator*, and cure rates are similar to those reported from the United Republic of Tanzania. What is of concern, however, is the fact that reductions in egg count were substantially worse than expected on the basis of current data from around the world (Bennett & Guyatt, 2000).

Table 4. Comparison of efficacy of different anthelminthics against *Necator* in Mali[a]

Treatment	Cure rate (%)	Change in mean eggs/g faeces (%)
Placebo	16.7	+38.9
Pyrantel	37.8	–45.5
Mebendazole	51.4	–68.5
Albendazole	83.8	–97.7

[a] Data from Sacko et al. (1999).

These two studies illustrate that, while resistance is probably rare at present, its development is a possibility, making it important that expansion of mass administration is paralleled by surveillance. Surveillance could take the form of efficacy studies in sentinel communities (it is essential to use the same community to minimize variability); alternatively, genetic markers for resistance may be useful for monitoring. It has also been suggested that changing the principal drug every year or two would reduce the risk of resistance. However, this will become a realistic proposal only when more alternative compounds are available.

New drugs on the block?

Since the approval of albendazole for the treatment of intestinal helminth infections in 1981, efforts to develop new drugs have been limited. Whether this stems from a lack of interest, a belief that the drugs currently available are adequate, or a lack of suitable compounds, there is now a realization that new products are needed in case resistance, such as seen in the veterinary field, develops in human helminths. As ever-larger mass control programmes are established, the risk of resistance grows. The following paragraphs summarize current work in the area of new drug development.

Nitazoxanide, a nitrothiazole anthelminthic, was first described by Rossignol & Cavier (1975) but has only recently been licensed for use in humans. Unusually for an anthelminthic, this compound was not originally developed for veterinary use. It is reported to have good activity not only against intestinal nematodes but also against *Entamoeba histolytica* and—like albendazole—*Giardia intestinalis.* Antiprotozoal activity is not unexpected in a compound closely related to metronidazole but an anthelminthic efficacy reportedly in excess of 90% for all common helminths is more surprising (Abaza et al., 1998). Nitazoxanide appears to be a promising therapeutic compound for individual treatment but will have limited utility for community treatment since multiple doses are required.

Oxibendazole, an albendazole analogue with oxygen substituted for sulfur, has long been used in veterinary medicine and is now under development for use against intestinal helminths in humans. Early evidence suggests that it is highly effective against *Ascaris*, at least as effective as albendazole in *Necator* infections, and more effective than albendazole for single-dose treatment of *Trichuris* infections (see Figures 2 and 3). The activity against *Trichuris* may be due to the apparent poor absorption of oxibendazole from the gastrointestinal tract, which results in very limited systemic exposure. Unlike most benzimidazoles, oxibendazole does not appear to be teratogenic in animals, and may therefore eventually be considered to be "safer" than other compounds for the management of hookworm anaemia during pregnancy.

Ivermectin is best known as a treatment for systemic helminth infections such as onchocerciasis and lymphatic filariasis. In veterinary medicine, it is normally

Figure 2. Cure rates in a dose-ranging study with oxibendazole

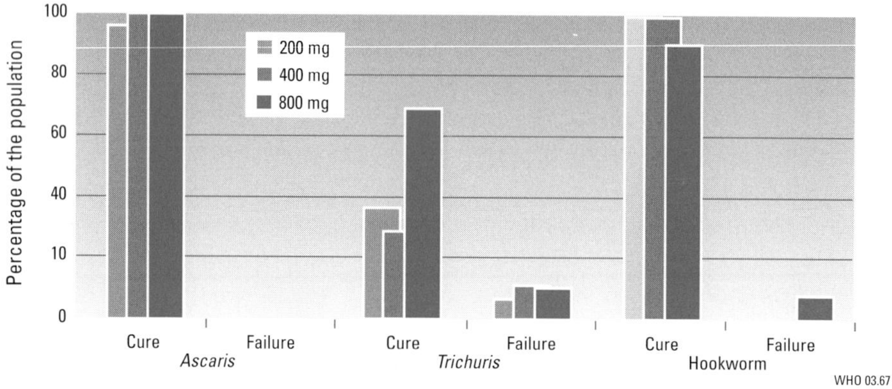

Figure 3. Efficacy of oxibendazole, single 400-mg dose, against *Ascaris, Necator,* and *Strongyloides*

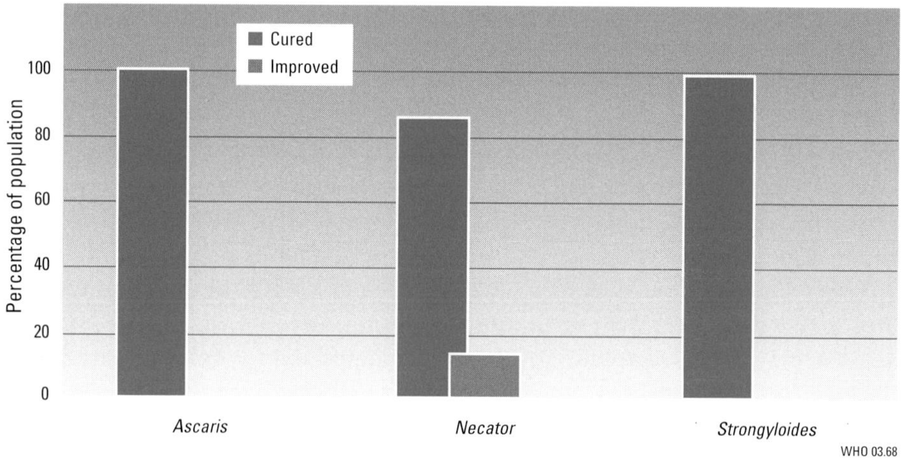

used for the treatment of intestinal infections, and recent data have shown single doses to be particularly effective in *Strongyloides* infections (Marti et al., 1996). This study also notes the effects of ivermectin against other intestinal nematodes, the almost total lack of effect in hookworm infections reported from a number of other studies, and the poor efficacy against *Trichuris.* Thus, while it may be useful for treating diagnosed infections where the infecting species is known, ivermectin—active against *Strongyloides stercoralis* and *Ascaris lumbricoides*—has little value for community treatment of soil-transmitted helminth infections. Moreover, since ivermectin is used in community treatment campaigns for the elimination of onchocerciasis in Africa, it might be expected that *Ascaris* would be reduced or eliminated in campaign areas, to be replaced by hookworm and *Trichuris.*

The Global Alliance to Eliminate Lymphatic Filariasis, established in 2000, also uses ivermectin to reduce circulating microfilaria but treatment is combined with albendazole. This approach means that hookworm is not specifically selected because of the broad activity spectrum of albendazole (see Figure 4); additionally there is a synergistic effect against *Trichuris* with much increased efficacy (Beach et al., 1999; M. Ismail, personal communication, 1999). It emphasizes the benefits of combined drug therapy (also exemplified by the older combination of pyrantel and oxantel), which will not only provide better cover and efficacy, but will also perhaps reduce the likelihood of resistance developing during large-scale drug distribution programmes.

There are two other compounds that may find a place in the treatment of soil-transmitted helminths. The first is triclabendazole, an unusual benzimidazole carbamate that seems to be well absorbed and may have activity against helminths; at present it is used principally to treat fascioliasis. The second is moxidectin, which has been shown in animals to be an effective replacement for ivermectin. It is currently under discussion as a filaricide, although its spectrum of activity may indicate usefulness in the management of soil-transmitted nematodes.

Figure 4. Efficacy of albendazole and ivermectin, alone and in combination, against helminths[a]

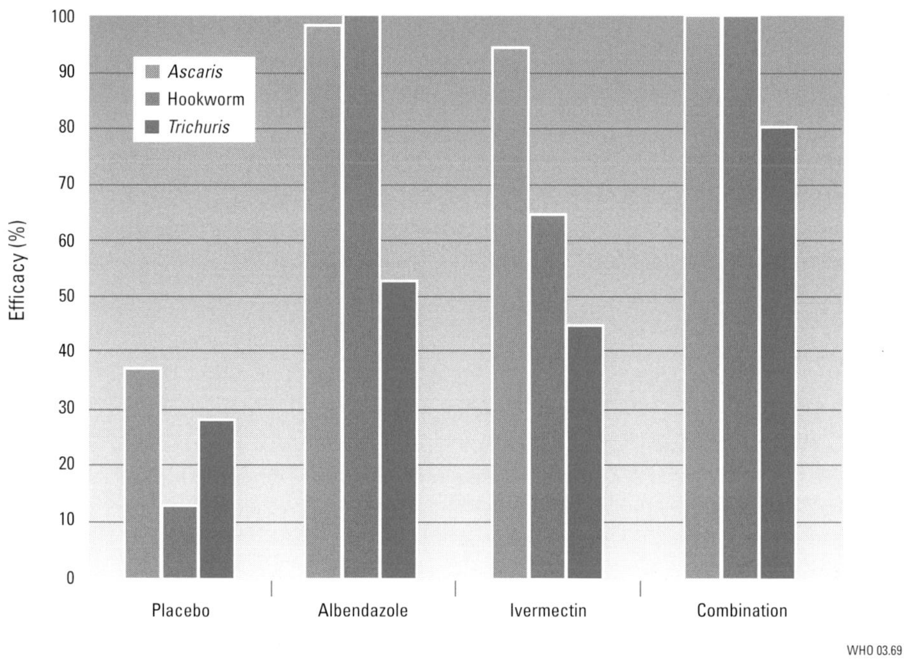

[a] Derived from Beach et al., 1999.

Conclusion

The anthelminthics in use today are generally effective and can be used for both individual and community treatment. There is room for improvement, however, especially in the treatment of *Trichuris* infections. Generic products can be useful provided that they are of adequate quality; otherwise, they can be a drain on financial resources and could encourage the development of resistance. Currently, there is little evidence of resistance developing in human helminths, although experience in animals suggests that continued vigilance is needed. The development of resistance would create a pressing need for new anthelminthics: a few are being developed but further efforts should be encouraged to ensure that novel products will be available in the future.

References

Abaza H et al. (1998) Nitazoxanide in the treatment of patients with intestinal protozoan and helminthic infections: a report on 546 patients in Egypt. *Current Therapeutic Research*, 59:116–121.

Albonico et al. (1994) A randomised controlled trial comparing mebendazole and albendazole against *Ascaris*, *Trichuris* and hookworm infections. *Transactions of the Royal Society of Tropical Medicine and Hygiene*, 88:585–589.

Amato-Neto V, Moreira AAB, Campos R (1983) Treatment of ankylostomiasis with albendazole. *Revista Instituto de Medicina Tropical de Sao Paulo*, 25:42–46.

Bartoloni A et al. (1993) Comparative efficacy of a single 400 mg dose of albendazole or mebendazole in the treatment of nematode infections in school children. *Tropical and Geographic Medicine*, 45:114–116.

Beach JM et al. (1999) Assessment of combined ivermectin and albendazole for treatment of intestinal helminth and *Wuchereria bancrofti* infections in Haitian school children. *American Journal of Tropical Medicine and Hygiene*, 60:479–486.

Bennett A, Guyatt H (2000) Reducing intestinal nematode infection: efficacy of albendazole and mebendazole. *Parasitology Today*, 16:71–74.

Bundy DAP (1997) This wormy world: then and now. *Parasitology Today*, 13:407–408.

Chandler AC (1929) *Hookworm disease*. New York, Macmillan.

Cruz O (1983) Comparative evaluation of the effect of single doses of mebendazole and albendazole. *Investigaciones Medica Internacional*, 10:404–406.

De Clercq D et al. (1997) Failure of mebendazole in the treatment of human hookworm infections in the Southern Region of Mali. *Annales de la Société Belge de Médecine Tropicale*, 75:191–199.

Der Simonian J, Laird R (1986) Meta-analysis in clinical trials. *Controlled Clinical Trials*, 7:177–188.

Galia E, Horton J, Dressman JB (1999) Albendazole generics—a comparative in vitro study. *Pharmaceutical Research*, 16:1871–1875.

Garcia EG (1976) Treatment for trichuriasis with oxantel. *American Journal of Tropical Medicine and Hygiene*, 25:914–915.

Holzer BR, Frey FJ (1987) Differential efficacy of mebendazole and albendazole against *Necator americanus* but not for *Trichuris trichiura* infestations. *European Journal of Clinical Pharmacology*, 32:635–637.

Ioli A et al. (1982) Albendazole in treatment of intestinal helminthiasis, study on 140 patients. *Rivista Parasitologica*, 4:293–295.

Janssens PC (1986) Chemotherapy of gastrointestinal nematodiases in man. In: Van den Bossche H, Thienpont D, Janssens PG, eds. *Chemotherapy of gastrointestinal helminths*. Berlin, Springer-Verlag:183–486.

Jonksuksuntigul P et al. (1993) A comparative study on the efficacy of albendazole and mebendazole in the treatment of ascariasis, hookworm infections and trichuriasis. *Southeast Asian Journal of Tropical Medicine and Public Health*, 24:724–729.

Mabey R (1986) *Flora britannica*. London, Sinclair Stevenson.

Marti H et al. (1996) A comparative trial of a single dose of ivermectin versus three days of albendazole for treatment of *Strongyloides stercoralis* and other soil-transmitted infections in children. *American Journal of Tropical Medicine and Hygiene*, 55:477–481.

Pugh RN, Teesdale CH, Burnham MM (1989) Albendazole in children with hookworm infections. *Annals of Tropical Medicine and Parasitology*, 5:133–136.

Reynoldson JA et al. (1997) Failure of pyrantel in the treatment of human hookworm (*Ancylostoma duodenale*) in the Kimberley region of north west Australia. *Acta Tropica*, 68:301–312.

Reynoldson JA et al. (1998) Efficacy of albendazole against *Giardia* and hookworm in a remote Aboriginal community in the north of Western Australia. *Acta Tropica*, 71:27–44.

Rossignol JF, Cavier R (1975) New derivative of 2-benzamido-5-nitrothiazoles. *Chemical Abstracts*, 83:28216n.

Rockefeller Foundation International Health Board (1922) *Publication 11: Bibliography of hookworm disease*. Baltimore, MD, Williams & Wilkins.

Sacko M et al. (1999) Comparison of the efficacy of mebendazole, albendazole, and pyrantel in the treatment of human hookworm infections in the Southern Region of Mali, West Africa. *Transactions of the Royal Society of Tropical Medicine and Hygiene*, 93:195–203.

Sinniah B, Chew PI, Sobramaniam K (1990) A comparative trial of albendazole, mebendazole, pyrantel pamoate and oxantel/pyrantel pamoate against soil-transmitted helminthiases in school children. *Tropical Biomedicine*, 7:129–134.

Stoll NR (1947) This wormy world. *Journal of Parasitology*, 33:1–18.

Wang BR et al. (1987) Comparative efficacy of thienpydin, pyrantel pamoate, mebendazole and albendazole in treating ascariasis and enterobiasis. *Chinese Medical Journal*, 100:928–930.

Effectiveness of selective treatment in the control of soil-transmitted helminthiases

Hoang Thi Kim, Nguyen Van De, Nguyen Van Tien, Nguyen Thi Dang, Pham Thi Tuyet, Nguyen Thi Mai, Hoang Van Tan, Nguyen Thi Duc, Nguyen Thi Viet Hoa[1]

Department of Helminthology
National Institute of Malariology, Parasitology and Entomology
Tu Liem, Hanoi, Viet Nam

Introduction

Soil-transmitted helminthiases are the most widespread diseases throughout the world, especially in tropical and subtropical countries. The diseases cause high morbidity and mortality, contribute to malnutrition, vitamin A deficiency, and anaemia, and have a negative impact on health generally and on the development of children (Pawlowski, Schad & Stott, 1991; Stephenson, 1987).

In northern Viet Nam, Kim (1987) estimated the prevalence of *Ascaris lumbricoides* infection as 85–95%, of *Trichuris trichiura* infection as 58–69%, and of hookworm infection as 30–60%.[2] In the south of the country, the prevalence of *A. lumbricoides* and *T. trichiura* infections was lower, while that of hookworm infection was similar. Multiple infections were common, with a prevalence of 50–89%. Children harboured both the highest prevalence and the highest intensity of *A. lumbricoides* infection; because they commonly defecate indiscriminately in the open, children are also a major cause of contamination of the environment.

The aim of the study reported here was to evaluate the efficacy of interventions designed to reduce infection intensity—and thereby to reduce the contamination of the environment, transmission of infection, and the adverse health effects and sequelae of helminthiases, and to improve children's overall health.

Between 1979 and 1982, a comparative study of the effect of mass treatment and treatment targeted to children was carried out in the Philippines; four treatments were given annually for three consecutive years. The results showed that

[1] Nguyen Thi Viet Hoa, Department of Helminthology, National Institute of Malariology, Parasitology and Entomology (NIMPE), Tu Liem, Hanoi, Viet Nam
Tel: +84 4 854 3035, Fax: +84 4 854 3015, E-mail: ntvhoa@fpt.vn

[2] Kim HT (1987) Some characteristics of epidemiology and control of Ascariasis in Vietnam. *National workshop on control of some important helminthiases in Vietnam. Hanoi, 15–21 October 1987.*

the two methods effected similar reductions in the prevalence of *A. lumbricoides*, *T. trichiura*, and hookworm infection (Cabrera & Cruz, 1983). The targeted approach was chosen for use in the study in Viet Nam: it was cheaper than the mass approach and thus more feasible given the financial resources available.

Method

Study site

The study was conducted in a rural hamlet, Ha Cau commune, Ha Dong town, located in Ha Tay province in northern Viet Nam. The area covers 36 hectares and, at the time of the study, had a population of 1630 adults (820 men, 810 women) and 466 children under 15 years of age. The people grow rice and vegetables; the use of fresh human exreta for fertilizing crops facilitates transmission of helminth infection. Hygienic conditions are poor, with only 7% of the community having access to adequate sanitation.

Target group and treatment

More than 90% of the children under 15 years of age received all courses of treatment; only those suffering from acute diseases were excluded from treatment. Three single-dose treatments with pyrantel pamoate (supplied as 125-mg tablets), at a dose of 10 mg/kg, were given annually for three consecutive years.

All children and adults were examined and then treated in the hamlet.

Follow-up of side-effects—monitoring and evaluation of results

Mass examination of stool samples (100 from children, 100 from adults) by Kato and Kato–Katz methods was carried out before treatment to determine the prevalence and intensity of infection (see Tables 1 and 2). Forty soil samples were examined (Romanenko method) before and after treatment annually for three years to assess the contamination of the environment with helminth eggs and to examine the relationship between contamination and the results of treatment. Stools from adults and children were examined annually, as follows, to assess the effect of the targeted treatment:

Table 1. Pretreatment prevalence rate of soil-transmitted helminthiases in children and adults

Age group (years)	No. of subjects examined	Worm infection		*A. lumbricoides*		*T. trichiura*		Hookworm	
		n	%	*n*	%	*n*	%	*n*	%
<15	225	208	92.4	188	83.6	174	77.3	28	12.1
≥15	329	319	96.9	295	89.6	281	85.4	91	27.7
Total	554	527	95.1	483	87.2	455	82.1	117	21.1

Table 2. Pretreatment intensity of soil-transmitted helminthiases in children and adults

Age group (years)	Infection intensity (eggs/gram faeces)		
	A. lumbricoides	*T. trichiura*	Hookworm
<15	8199.9 (0–55 500)	264.3 (0–5550)	16.4 (0–462.5)
≥15	5815 (0–42 550)	271.8 (0–5550)	42.7 (0–1480)

Table 3. Treatment efficacy of pyrantel pamoate

Species	No. of patients followed up	CR[a] (%)	ERR[a] (%)
Ascaris	91	92.3	96.0
Trichuris	93	23.7	32.5
Hookworm	13	100.0	100.0

[a] CR = cure rate, ERR = egg reduction rate.

Table 4. Reinfection 4 months after the first course of treatment

Species	No. of patients followed up	No. of patients reinfected	% reinfection
Ascaris	59	58	98.3
Trichuris	2218	81.8	81.8
Hookworm	13	0	0

Before treatment:	554 persons
After the first year's treatment:	518 persons
After the second year's treatment:	338 persons
After the third year's treatment:	407 persons

Results

Prevalence and intensity of infection

Results are summarized in Tables 3–6; see also Figures 1–4. The overall prevalence of *A. lumbricoides* infection in the community after the first, second, and third years of treatment declined slowly, by a total of about 10% (see Table 5); however, the differences were highly statistically significant ($P < 0.001$). In children, prevalence declined more than in adults after three years of treatment, especially after the first year ($P < 0.001$) compared with the second and third years. The reason for this slow reduction might be poor environmental hygiene: if the environment were still contaminated with helminth eggs, there would have

Figure 1. Prevalence rates of soil-transmitted helminthiases in children before and after selective treatment

a) *Ascaris lumbricoides*

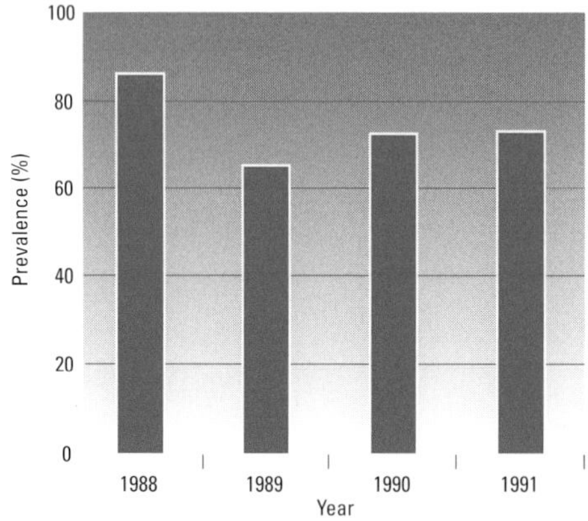

b) *Trichuris trichiura*

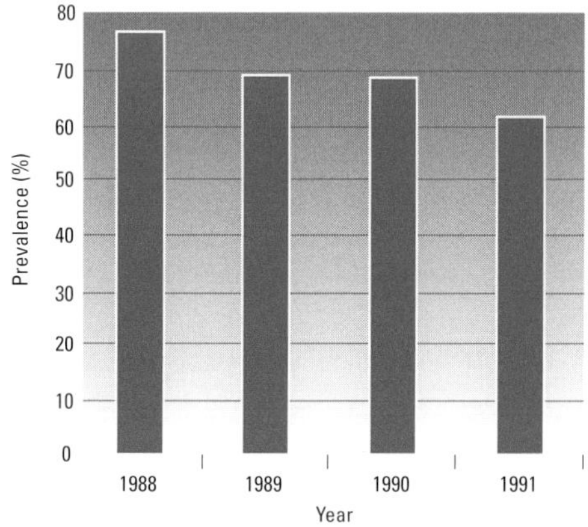

c) Hookworm

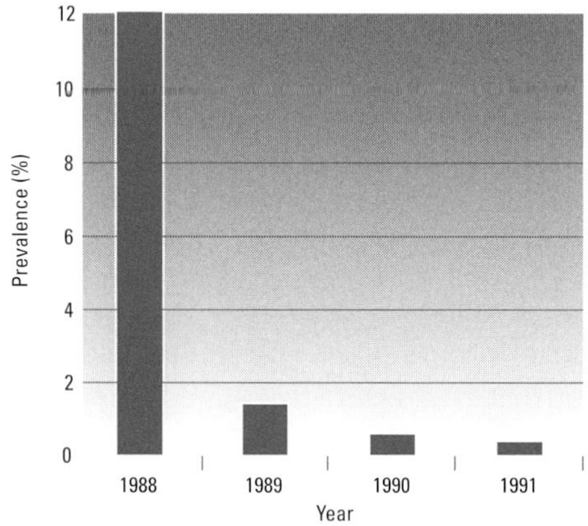

WHO 03.80

Figure 2. Prevalence rates of soil-transmitted helminthiases in adults before and after treatment

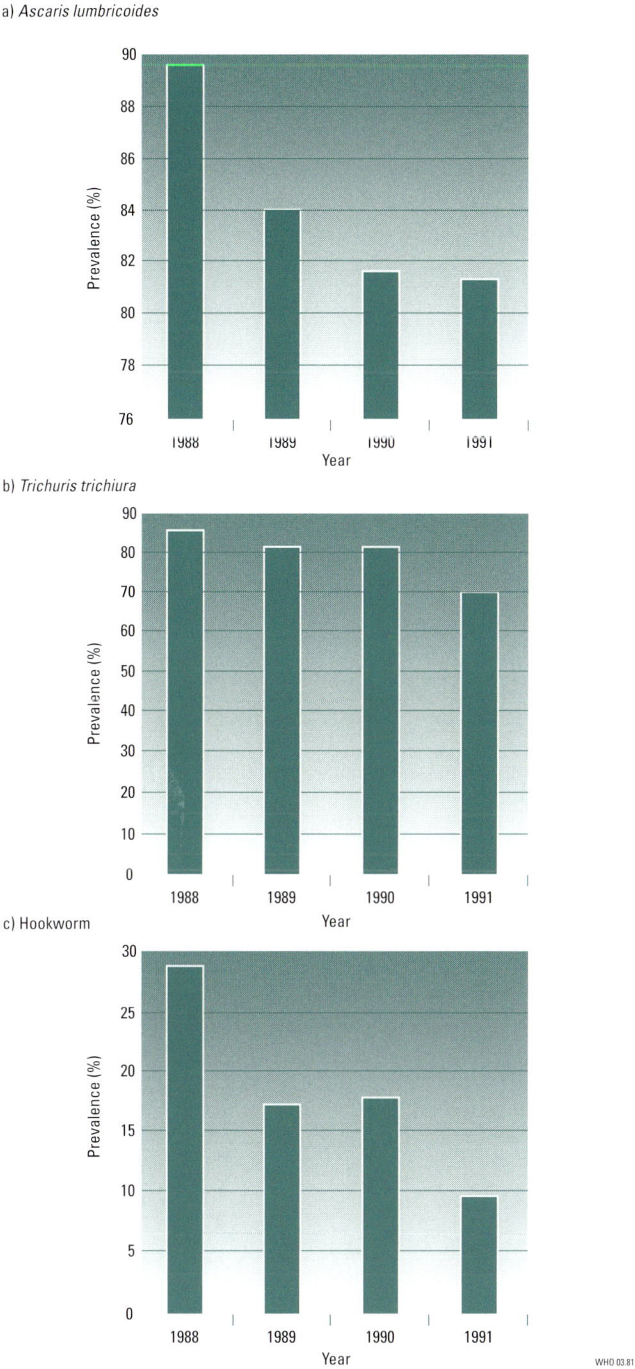

Figure 3. Infection intensity of soil-transmitted helminthiases in children before and after selective treatments

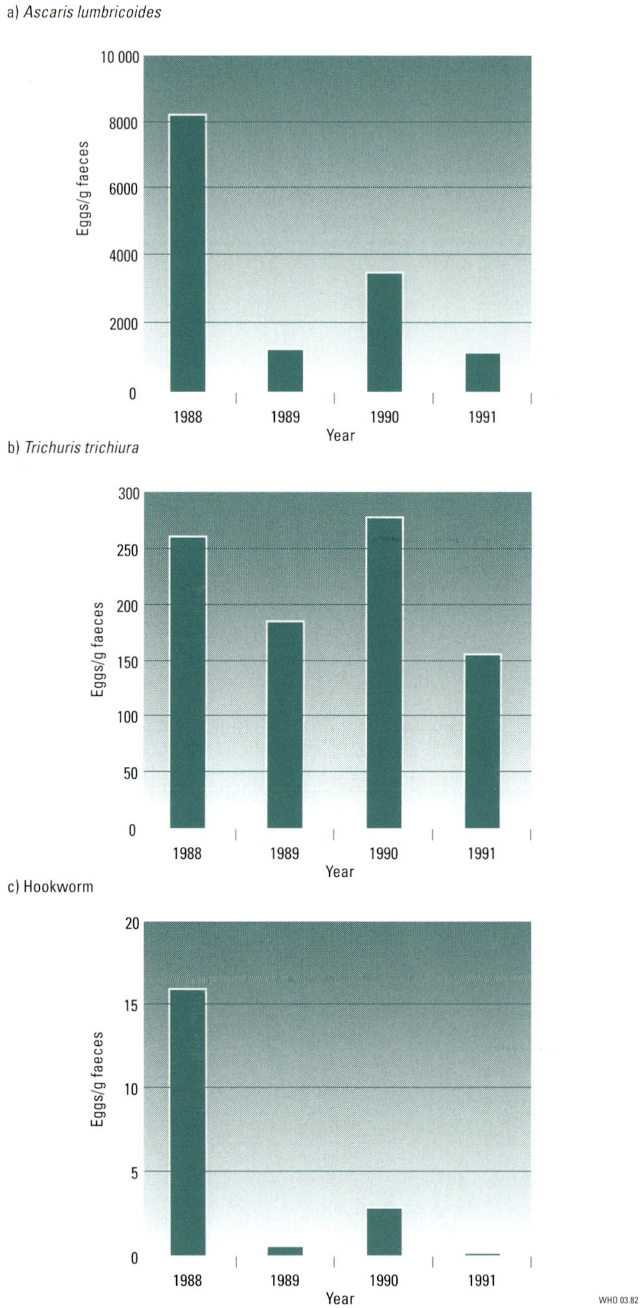

Figure 4. Infection intensity of soil-transmitted helminthiases in adults before and after selective treatments

a) *Ascaris lumbricoides*

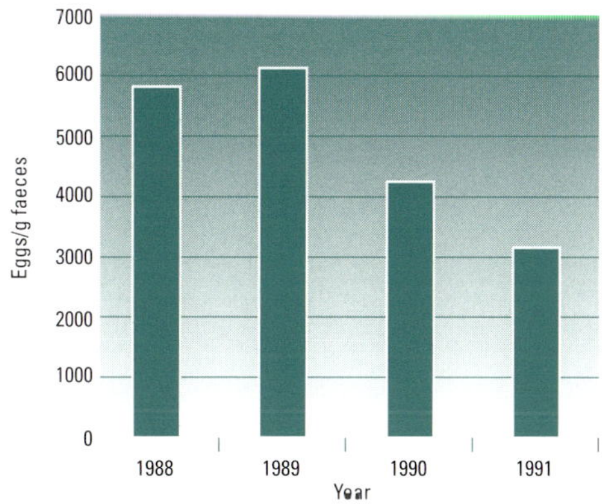

b) *Trichuris trichiura*

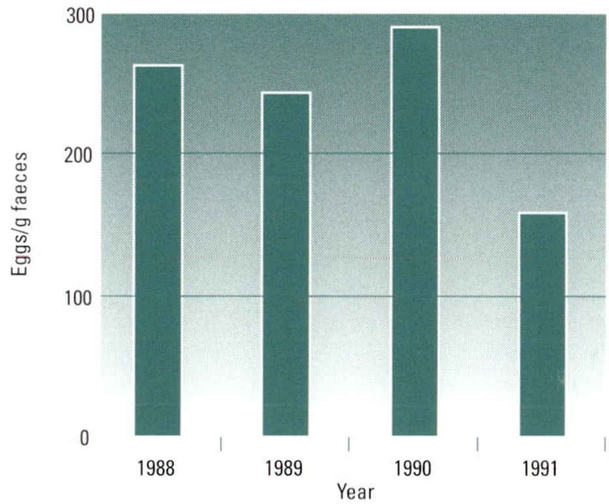

c) Hookworm

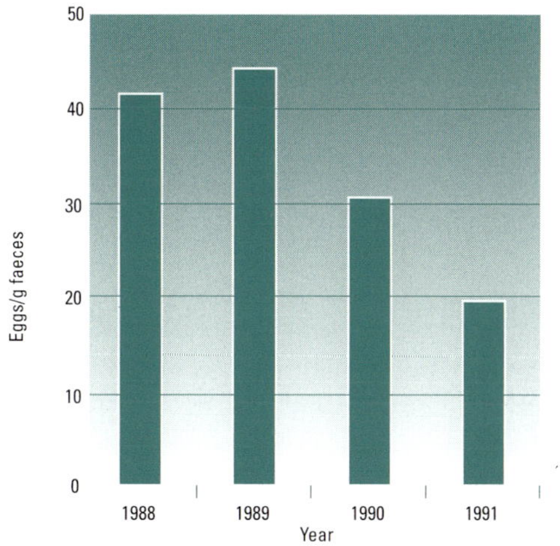

WHO 03.83

Table 5. Prevalence rates of soil-transmitted helminthiases in children and adults before and after selective treatment

Time examined	Children				Adults				Total			
	No. examined	% infected			No. examined	% infected			No. examined	% infected		
		A. lumbricoides	*T. trichiura*	Hookworm		*A. lumbricoides*	*Trichuris*	Hookworm		*A. lumbricoides*	Trichuris	Hookworm
Pretreatment	225	83.6	77.3	12.1	329	89.6	85.4	27.7	554	87.2	82.1	21.1
After 1st year of treatment	193	63.2	69.4	1.5	325	83.9	81.2	17.2	518	76.1	76.8	11.4
After 2nd year of treatment	149	70.3	69.1	0.7	189	81.5	80.4	18.0	338	76.6	75.4	10.3
After 3rd year of treatment	156	70.5	62.8	0.6	251	81.1	69.3	9.6	407	77.1	66.8	6.1

Table 6. Infection intensity of soil-transmitted helminthiases in children and adults before and after selective treatment

Time examined	Infection intensity (eggs/gram faeces)					
	A. lumbricoides		T. trichiura		Hookworm	
	Children	Adults	Children	Adults	Children	Adults
Pretreatment	8199.9	5845.0	264.3	271.8	16.4	42.7
After 1st year of treatment	1366.5	6112.6	183.3	256.9	0.5	45.5
% egg reduction	83.3	—	30.7	5.5	96.9	—
After 2nd year of treatment	3515.8	4231.6	276.4	294.6	2.5	31.4
% egg reduction	57.1	27.6	—	—	84.8	26.5
After 3rd year of treatment	1031.6	3116.8	152.1	157.7	0.1	19.6
% egg reduction	87.4	46.7	42.4	41.9	99.6	45.9

been both reinfections and new infections. The intensity of *A. lumbricoides* infection (see Table 6), by contrast, decreased markedly and rapidly in both adults and children (egg reduction rate in adults was 27.6–46.7% and in children 57.1–87.4%).

For *T. trichiura*, the prevalence and intensity of infection declined, but the pattern was inconsistent. It is possible that pyrantel was less effective against *T. trichiura*.

The overall prevalence of hookworm infection in the community declined markedly, from 21.1% to 6.1% after three years of treatment. In children, prevalence of infection fell from 12.1% to 0.6%; in adults, prevalence was reduced from 27.7% to 9.6%.

The effect of this selective treatment on *A. lumbricoides* and hookworm infection was similar to that of mass treatment in the past, but the targeted treatment was cheaper (Kim et al., 1991). In comparing results with the study of targeted treatment of children in the Philippines (Cabrera & Cruz, 1983), it is clear that our results were lower, especially for *A. lumbricoides*: after 3 years of treatment in the Philippines, *A. lumbricoides* prevalence declined from 68.4% to 11.4%, and hookworm infection from 29.7% to 6.3%. It may be that environmental hygiene was better in the Philippines, so that there was less reinfection.

Environmental contamination with helminth eggs

Targeted treatment had no significant effect in reducing environmental contamination with *A. lumbricoides* eggs ($P > 0.05$), although the egg burden had decreased markedly after 3 years' treatment, from 70.6 eggs/100 g soil to 28.5 eggs/100 g. On the other hand, the percentage of non-viable eggs in the soil increased (see Tables 7 and 8). After 3 years' treatment, most of the *A. lumbricoides* eggs in children's stool samples were unfertilized, showing that treatment was not only reducing the prevalence and intensity of infection but was also

Table 7. Ascaris lumbricoides egg contamination in soil before and after selective treatment

Time examined	No. of soil samples examined	No. positive	% positive	Average no. eggs/100 g soil	% non-viable eggs
Pretreatment	40	40	100	70.6	18.8
After first year of treatment	40	38	95.0	60.1	24.8
After second year of treatment	40	37	92.5	40.4	38.5
After third year of treatment	40	37	92.5	28.5	40.3

Table 8. Trichuris trichiura egg contamination in soil before and after selective treatment

Time examined	No. of soil samples examined	No. positive	% positive	Average no. eggs/100 g soil	% non-viable eggs
Pretreatment	40	35	87.5	7.8	2.0
After first year of treatment	40	33	82.5	5.3	9.1
After second year of treatment	40	32	80.0	6.8	22.9
After third year of treatment	40	33	82.5	4.7	22.2

affecting the fertility of helminth eggs. Unfertilized eggs do not present a health hazard.

Environmental contamination with *T. trichiura* eggs after three years of treatment had not declined significantly compared with pretreatment levels ($P > 0.05$), and comparison with the effect on *A. lumbricoides* eggs showed no clear-cut reduction in egg viability. The explanation for this is probably the lower efficacy of pyrantel against *Trichuris*.

Conclusion

After three years' selective treatment with pyrantel pamoate, the prevalence of *A. lumbricoides* infection had declined by only about 10%. However, the objective of reducing infection intensity in the community in general—and in children in particular—was achieved, with the result that environmental contamination with helminth eggs was reduced, as well as transmission of infection. Both prevalence and intensity of hookworm infection decreased sharply in children and moderately in adults. The efficacy of the treatment against *T. trichiura* infection was unclear.

The results of this study indicated that selective treatment of children could be widely applied. The anthelminthics and duration of treatment (2–4 courses per year) will be chosen according to the economic situation and worm infection characteristics of each region. At present, priority should be given to rural children with high rates of worm infection.

References

Cabrera BD, Cruz I (1983) A comparative study on the effect of mass treatment of the entire community and selective treatment of children on the total prevalence of soil-transmitted helminthiases in two communities in Mindoro, Philippines. In: *Collected papers on the control of soil-transmitted helminthiases*, 2:226–287. Tokyo, Asian Parasite Control Organization.

Kim HT et al. (1991) Evaluation of effectiveness of some anthelminthics used in Vietnam. Scientific works (1986–1990). Vol. II: Parasitology and Entomology:38–45.

Pawlowski ZS, Schad GA, Stott GJ (1991) *Hookworm infection and anaemia: approaches to prevention and control.* Geneva, World Health Organization.

Stephenson LS (1987) Human malnutrition and helminths. In: Stephenson LS, Holland C, eds. *The impact of helminth infections on human nutrition: schistosomes and soil-transmitted helminths.* London, Taylor & Francis:1–20.

PART IV

Experience from national control programmes

Experience from national control programmes

A wealth of knowledge has been acquired about how to plan, implement, and sustain programmes designed to control and reduce the burden of disease due to helminth infections. Much of this experience is published and is freely available in the public domain. There are three important features embedded in this body of experience. First, it helps health professionals assigned to the task of helminth control in coping with the intricate details of setting up programmes, from advocacy to drug delivery, from technical training to data analysis. Second, it identifies pitfalls and problems and suggests how these might be avoided—or managed when encountered. Third, it is a source of encouragement; those embarking on helminth control can be confident that success can be achieved. There is now no doubt that controlling morbidity due to helminthiasis is an attainable goal.

The Editors

Control of schistosomiasis due to Schistosoma mekongi *in Khong District, 1989–1999*

T. Khamkeo,[1] K. Pholsena
Hygiene and Preventive Department, Ministry of Public Health
Vientiane, Lao People's Democratic Republic

Introduction

The Lao People's Democratic Republic is located in south-east Asia. Its geographical and climatic characteristics, combined with a lack of development, contribute to the incidence and prevalence of communicable diseases. Intestinal helminth infections affect approximately 90% of the population, and double or even triple infections are common. The most frequent nematode infections encountered are ascariasis, trichuriasis, ancylostomiasis, and strongyloidiasis. Infections with cestodes, including *Taenia solium*, *T. saginata*, and *Hymenolepis nana*, are less common. Trematode diseases are important because they depend essentially on the lifestyle and food habits of the population, who are either unaware of preventive measures or insufficiently motivated to adopt them. Cases of paragonimiasis and infection with *Fasciolopsis buski* and *Fasciola hepatica* are relatively rare among humans. The most prevalent trematode is *Opisthorchis viverrini*; infection is caused by eating raw fish in various forms and occurs throughout the country. There is also a focus of schistosomiasis due to *Schistosoma mekongi* in Khong district, which is located in the southern province of Champasak on the border with Cambodia.

Khong district covers 2360 km^2. Its population of approximately 65 000 is distributed among 13 communes and 131 villages and depends on rice cultivation, fishing, and tourism. The Mekong, which flows through the district, contains thousands of islands, the largest of which is Khong (20 km long and 8–10 km wide), the district capital, more commonly known as Muong Khong ("the town of Khong").

Human parasite infection depends on contact with the water of the Mekong, a rich source of food for the population around which life has been organized for many centuries. The Mekong is a rich source of food for the population of Khong district, and life has been organized around the river for centuries. However, the river water is also the source of human parasite infections. For generations, every island community has had its "fat-bellied people" yet remained unaware of the

[1] Dr T. Khamkeo, Hygiene and Preventive Department, Ministry of Public Health, Vientiane, Lao People's Democratic Republic
Tel/fax: +856 2121 4010

parasites that cause the affliction. Controlling *S. mekongi* schistosomiasis in order to help protect public health thus became one of the most urgent tasks facing the Ministry of Health. With the help of WHO's Regional Office for the Western Pacific, measures to control *S. mekongi* infection began in Khong district in November 1988 and continued for almost seven years, to December 1995. Thereafter, control efforts continued with the assistance of the German Pharma Health Fund and was extended to seven other riverside districts: Moulapamok, Soukouma, Pathouphone, Champasak, Paksé, PhonThong, and Xanasomboun. This assistance officially ceased at the end of 1999.

Surveys and feasibility studies

Control of schistosomiasis due to *S. mekongi* in Khong district involved a number of surveys and feasibility studies to locate the sites of transmission, since only fragmentary information was available.

Sociological survey (April to May 1989)

A sociological survey was conducted by questionnaire among a sample of 100 families in the neighbourhood of Chomthong, one of the towns in the district. The survey was designed to reveal both aspects of the population's behaviour that were likely to have an influence on the spread of schistosomiasis and popular beliefs about the disease. The survey found that 100% of the families interviewed bathed in the Mekong: 97% of them bathed exclusively in the river, while 3% also occasionally used well-water. All families used water from the Mekong for domestic purposes; 2% also used well-water. The survey also revealed that all the families fished in the Mekong; this inevitably entailed wading in the water, regardless of the fishing technique used. Forty percent of the families said that they had never used latrines; 8% of the families who had used latrines said that they also defecated outdoors occasionally. In addition, 38% of the families said that their children usually defecated in the Mekong.

When a range of symptoms were described in order to identify popular beliefs about schistosomiasis, 1% of the families said that they were completely ignorant about the disease, 92% that they had heard of it, and 7% that they were familiar with it. However, none of the families interviewed knew of any link between the disease and water snails, which explained their lack of concern about the health risk of frequent contact with the Mekong. Clearly, health education is of paramount importance, and people have first and foremost to be educated about the link between the disease and the snails. Before control measures were introduced, the expression *phayad hoy Muang Khong* (literally, "the mollusc disease in Khong district") was adopted to publicize the notion of this link and, by association, to establish another notion, the relationship between the disease and the waters of the Mekong.

The parasitological survey (November 1988 to December 1990)

The parasitological survey was conducted in 30 villages located in 6 communes. It involved examination of fresh stool samples, plus either physical examination of the liver alone (25 villages, 5 communes) or examination of both liver and spleen (8 villages, 3 communes). The target population was primary- and secondary-school children aged 4–15 years.

The stool specimens were examined using the Kato–Katz thick smear technique. The liver was deemed enlarged if it was palpable either below the right costal margin on the midclavicular line, or 3 mm below the apex of the xiphisternum. The spleen was evaluated using Hacket's method. A total of 2249 Kato–Katz examinations and 2585 liver examinations (1225 of them in association with spleen examinations) were completed.

Eggs of *S. mekongi* were found in people from all of the villages investigated except two—Houa Sène (Muang Sène commune), where the bed of the Mekong is made up solely of sand and does not support molluscs, and Set Nam Om (Kadane commune), located more than 6 km from the river. The average prevalence of infection was 42.2% (993 egg-carriers among 2249 persons examined). Prevalence varied from village to village, ranging from a minimum of 5.4% in Houa Khong Tay (Houa Khong commune) to a maximum of 98.2% in Ban Na (Kang Khong commune). Overall, prevalence tended to increase from Khong island to the Cambodian border. Egg counting in a number of villages gave an average epg (eggs/gram faeces) value varying from 31.2 to 136.6. There was an apparent relationship between this measure of infection intensity and the prevalence of *S. mekongi* infection (see Table 1). As expected, examination of fresh stool samples also revealed the presence of eggs of other helminths, such as *Ascaris*, *Ancylostoma*, *Trichuris*, *Strongyloides*, and *Opisthorchis*. Indeed, prevalence of *Opisthorchis* exceeded that of *S. mekongi*, with which it coexists in all the villages surveyed (Table 2).

Physical examination showed liver enlargement in 23.3% of the individuals examined (607/2585) and spleen enlargement in 7.5% (92/1225). However, there was no statistically significant relationship between *S. mekongi* infection status and organ enlargement. The high proportion of persons with enlarged organs in certain villages seemed to be related to infection with other parasites coexisting in the region, particularly *O. viverrini* and the malaria parasite. Further physical examinations of liver and spleen were therefore abandoned.

The parasitological survey also revealed the importance for transmission of *S. mekongi* of the location of a village. The risk of infection is low, or even non-existent, if the river bed near the village does not provide an environment suitable for *Neotricula* snails, such as rocky outcrops, or submerged branches or metal cans. This is the case for Houa Sène (Muang Sène commune) and for

Table 1. Relationship between prevalence and intensity of Schistosoma mekongi infection, 1989–1990

Villages surveyed	Prevalence (%)	Average intensity (eggs/gram faeces)
1989		
Ban Dong	38.3	40.5
Ban Houay	40.4	49.0
Ban Na	98.2	136.8
Hin Siu	15.3	43.6
Kang Khong	42.6	38.2
Sène Had Gnay	68.2	47.6
Sène Had Noy	20.7	45.5
Sène Lam	42.8	47.6
Sène Tay	8.2	57.8
Xieng Vang	21.1	31.2
1990		
Som Van Tok	72.5	72.1
Tha Kham	77.1	59.6
Tha Mak Hep	94.0	56.1

Table 2. Comparative prevalence of Schistosoma mekongi and Opisthorchis viverrini infection, 1989–1990

Villages surveyed	Prevalence of *S. mekongi* (%)	Prevalence of *O. viverrini* (%)
1989		
Ban Houay	40.4	42.1
Hin Siu	15.3	63.7
Sène Tay	8.2	63.0
1990		
Som Van Ok	76.9	41.5
Tha Kham	77.1	71.2
Tha Mak Hep	94.0	63.6

villages further away from the Mekong, such as Set Nam Om (Kadane commune), which is more than 6 km from the river. In both Houa Sène and Set Nam Om, all the stool slides examined were negative for *S. mekongi*.

Endemic schistosomiasis due to *S. mekongi* in Khong district was confirmed by the survey, which also revealed the importance of *O. viverrini* infection. The latter infection should be taken into account by the control programme, particularly since the efficacy of praziquantel against opisthorchiasis is well known.

The malacological survey (October 1990 to June 1991)

The malacological survey, conducted during the periods of high water (October–November 1990) and low water (May–June 1991), was completed by a trial involving the chemical molluscicide niclosamide and carried out on a 150 × 15 m stretch of water along the eastern side of Khong island. At the point furthest upstream in the area to be treated, 2 kg of niclosamide, wrapped in small bags, were introduced into the water. The trial was evaluated by calculating the population density (per 100 cm^2 surface of rock habitat) of the *Neotricula* collected at random in the trial area. The molluscs were examined microscopically and screened for larval forms of *S. mekongi* (crushed mollusc slides).

A considerable predominance of the gamma race of *Neotricula* was noted, with rare representatives of the alpha race and a complete absence of the beta race. The gamma race of *Neotricula* can therefore be considered to be the natural intermediate host of *S. mekongi* in this area.

During the high-water period, most of the *Neotricula* collected were young molluscs with 3.0 to 3.5 spires; they were found sticking to fragments of rock at a depth of 2–3 metres. During the low-water period, however, the specimens collected were mainly adults, with 4.0 to 5.5 spires, found in shallow water (10–60 mm deep), sticking to fragments of rock and submerged branches and cans. Transmission must therefore occur during the low-water period—more precisely, during April and May, when all the *Neotricula* collected had reached the adult stage.

The niclosamide trial failed to match expectations. Density of the *Neotricula* population was found to have increased 5 days after the introduction of the chemical into the trial area (see Table 3). This clearly shows the impracticability of localized application of the molluscicide when contact between the population and contaminated water is spread over long reaches of the river, tens of kilometres in length, and especially when the volume of flow is as high as that of the Mekong. This type of situation requires constant spraying—a costly undertaking in terms of material and logistic resources, in addition to malacology experts.

Efficacy of praziquantel

A coprological study, based on modification of *S. mekongi* egg excretion in the stools of schoolchildren, was undertaken to demonstrate the efficacy of prazi-

Table 3. Change in the population density of Neotricula during the chemical control trial

Distance from dispensing point	Population density of Neotricula (per 100 cm^2 rock surface)			
	before trial	after 1 day	after 5 days	after 9 days
10 metres	1.70	0.09	4.08	3.90
70 metres	12.90	0	2.10	11.0
150 metres	1.90	0	5.40	22.90

Table 4. Effect of single-dose praziquantel (40 mg/kg body weight) on excretion of Schistosoma mekongi eggs in stools (Sène Had Gnay)

Date of examination	Total no. examined	No. of positive results(%)	No. of negative results (%)
25 April 1989[a]	65	55 (84.6)	10 (15.4)
10 October 1989	65	10[b] (15.4)	55 (84.6)

[a] Praziquantel administered after initial examination.
[b] Of the 10 positive results, 8 were positive and 2 were negative on initial examination.

Table 5. Adverse effects of a single dose of praziquantel (40 mg/kg body weight) in 102 schoolchildren (Keng Koum, December 1990)

Adverse effects	No of pupils affected (%)
Headaches	54 (52.9)
Diarrhoea	43 (42.1)
Abdominal pain	20 (19.6)
Vomiting with diarrhoea	10 (9.8)
Vomiting without diarrhoea	7 (6.8)
Dry throat and cough	6 (5.9)
Fever with vomiting and/or diarrhoea	3 (2.9)
Fever without either vomiting or diarrhoea	2 (1.9)
Nausea	3 (2.9)
Itching and rash	3 (2.9)
Anorexia	3 (2.9)

quantel. Two examinations of fresh stools from 65 pupils at a school in Sène Had Gnay were undertaken—an initial examination, followed by administration of a single dose of praziquantel (40 mg/kg body weight) in April 1989 and a control examination some 6 months later in October. Results demonstrated the considerable efficacy of praziquantel against *S. mekongi* infection: prevalence was reduced from 84.6% on initial examination to 15.4% 6 months later (Table 4).

In a further study at a school in Keng Koum (Done Som commune), a group of pupils treated with praziquantel tolerated the drug well despite fairly frequent adverse effects such as vertigo, diarrhoea, vomiting, and abdominal pain. These adverse effects were generally manifest 1–2 hours after administration of the drug and regressed after 6–12 hours; exceptionally, effects persisted for up to 3 days (Table 5).

Control measures

Actual control measures against *S. mekongi* in Khong district rely essentially on mass chemotherapy with praziquantel, supported by health education.

Mass chemotherapy (April 1989 to December 1995)

Using praziquantel at a single dose of 40 mg/kg of body weight, and ensuring that recipients took the dose on the spot, mass treatment was applied to the entire population with the exception of:

— children under 2 years old
— pregnant women
— women breastfeeding babies under 6 months old
— people with acute illnesses
— people with neurological disorders and/or psychological disorders (e.g. epilepsy, chronic alcoholism)
— inhabitants of villages located more than 6 km from the Mekong.

Obliging people to take the praziquantel on the spot was considered to be a necessary precaution against their throwing away the drug, giving it to others, or keeping it with a view to selling it.

Two operational strategies were adopted successively—treatment of selected groups and universal treatment.

Under the selective strategy, the population to be treated in each village was determined by the prevalence of *S. mekongi* among the schoolchildren examined:

- If prevalence rate was 50%, the entire population aged 2–69 years was treated.
- If prevalence was between 25% and 50%, all children aged between 2 and 14 years were treated.
- If prevalence rate was below 25%, only children aged 2–14 years with *S. mekongi* eggs in their stools were treated.

This strategy meant that distribution of praziquantel depended on completion of a parasitological survey at each treatment site. Because of certain inconsistencies in both the progress of the surveys and the distribution of drugs (some villages were visited two or three times by the mobile teams), treatment of selected groups in all the target villages in the district took almost 2 years (April 1989 to December 1990). This prompted the adoption of a universal treatment approach in 1991.

Under the universal strategy, praziquantel (40 mg/kg of body weight) was given to the entire population aged over 4 years without prior stool examination; the dose was taken on the spot. Between August 1991 and December 1995, four praziquantel treatment cycles were completed, each requiring less than one month. If the earlier treatment of selected groups is included, a total of five cycles of treatment were completed over approximately 6 years (April 1989 to December 1995). Each cycle covered all the communes and from 105 to 129 villages (80–98% of the villages in those communes). Between 19 883 and 37 689 people were treated, i.e. 34–65% of the total population, or 46–88% of the target population (see Table 6). The sixth and last praziquantel treatment cycle in Khong

Table 6. Mass treatment cycles with praziquantel in Khong district (April 1989 – December 1995)

Cycle	Dates	Locations (% coverage)		Total no. of persons treated	Coverage as percentage		Praziquantel distributed (600-mg tablets)	
		Communes	Villages[a]		of total population[b]	of target population	Total no. of tablets	Average dose per person
I	Apr 1989–Dec 1990	13 (100%)	129 (98.5%)	37 689	65.0	88.0	93 177	2.5 tablets
II	Aug–Sep 1991	13 (100%)	120 (91.6%)	21 266	37.0	49.0	52 276	2.5 tablets
III	Feb–Mar 1993	13 (100%)	105 (81.1%)	21 642	37.0	50.0	52 851	2.4 tablets
IV	Aug–Sep 1993	13 (100%)	107 (81.6%)	20 615	36.0	48.0	52 479	2.5 tablets
V	Nov–Dec 1995	13 (100%)	108 (82.4%)	19 883	34.0	46.0	46 500	2.3 tablets

[a] A number of villages a long way from the Mekong were either not treated during the mass treatment cycle or were treated irregularly.
[b] Depending on the source, the population of Khong district varies from 58 000 to 68 000. The lower figure was chosen and the estimated percentage for whom treatment was contraindicated to be 26%.
[c] Treatment of selected groups.

Table 7. Mass treatment cycle with praziquantel in eight riverside districts (March– April 1998)

Cycle	Dates	Locations (% coverage)		Total no. of persons treated	Coverage as percentage		Praziquantel distributed (600-mg tablets)	
		Communes	Villages[a]		of total population[b]	of target population	Total no. of tablets	Average dose per person
VI	Mar–Apr 1998	8 (100%)	424 (60.8%)	129 585	48.8	52.2	329 668	2.5 tablets

district coincided with the first in the other seven riversides districts; it was administrated in March–April 1998 (Table 7).

Health education

Health education is a key element of the control strategy. Its purpose is to help the population to understand that schistosomiasis is attributable to individual behaviour in relation to the waters of the Mekong river and to the practice of indiscriminate defecation.

Health education was provided through workshops and training sessions organized on Khong island itself and through the distribution of illustrated posters and calendars (which were much sought after). The posters were often distributed immediately after meetings at which schistosomiasis due to *S. mekongi* was explained.

Conclusion

The programme to control schistosomiasis due to *S. mekongi* in Khong district lasted some 10 years (1989–1999). Overall, despite limited resources and a lack

Table 8. Change in prevalence rates of S. mekongi infection among schoolchildren in the villages monitored

Villages surveyed and monitored	Preliminary survey (1989–1990)	Prevalence after implementation of control measures		
		Jul–Aug 1993	Dec 1994	May–Jun 1999
Hang Som	92.3 (51/54)	0.0 (93/0)		0.0 (29/0)
Veune Som	72.5 (109/79)	0.0 (19/0)		0.0 (36/0)
Som van Tok		2.3 (44/1)		2.7 (37/1)
Tho La Thi	72.0 (57/41)	2.2 (46/1)		0.0 (25/0)
Keng Koum	62.8 (140/88)	8.7 (69/6)		0.0 (35/0)
Tha Pho	64.3 (56/36)	2.0 (49/1)		0.0 (25/0)
Tha Phao	60.5 (43/26)	2.6 (77/2)		0.0 (37/0)
Tha Mak Hep	94.0 (33/31)		0.0 (48/0)	5.0 (40/2)
Tha Kham	77.1 (118/91)		0.0 (77/0)	0.0 (46/0)
Som Van Ok	76.9 (65/50)		0.0 (48/0)	7.7 (39/3)
Sène Lam	42.8 (63/27)	0.0 (56/0)	0.0 (123/0)	0.0 (38/0)
Kang Khong	42.6 (54/23)	0.0 (96/0)	0.0 (152/0)	0.0 (60/0)
Xien Vang	21.1 (57/12)	0.0 (100/0)	0.0 (99/0)	0.0 (56/0)
Ban Dong	38.3 (107/41)		0.9 (117/1)	0.0 (99/0)
Sène Had Gnay	68.2 (151/103)	2.3 (44/1)	1.6 (63/1)	1.3 (76/1)
Sène Had Noy	20.7 (29/6)	0.0 (32/0)	0.0 (21/0)	0.0 (26/0)
Hin Siu	15.3 (137/21)	0.0 (63/0)	1.1 (87/1)	0.0 (57/0)
Sène Tay	8.2 (194/16)	0.0 (69/0)	0.0 (158/0)	2.8 (35/1)
Ban Na	98.2 (56/55)	0.0 (97/0)	0.0 (119/0)	0.0 (58/0)
Ban Houay	40.4 (156/63)	0.0 (57/0)	0.8 (122/1)	0.0 (58/0)
Khinak	41.4 (58/24)	6.3 (63/4)	0.0 (321/0)	0.0 (63/0)

of experience among the personnel, the programme has been largely successful. Coverage of the population by mass treatment was excellent. Health education campaigns had a substantial impact on the population, facilitating progressive behavioural change in terms of contact with the waters of the Mekong and use of latrines. Compared with levels found during the initial surveys, prevalence of *S. mekongi* infection in the 21 sentinel villages declined dramatically (Tables 8 and 9). Each successive survey during the 10-year programme revealed this reduction:

- In 1989 (21 villages surveyed), prevalence was estimated to be more than 50%.
- In 1993 (17 villages surveyed), prevalence had decreased to 1.5%.
- In 1999 (21 villages surveyed), prevalence was estimated to be 0.8%.

Table 9. Change in average prevalence of S. mekongi infection among schoolchildren in the villages monitored

	7 villages		14 villages		21 villages	
	Preliminary survey (1989–1990)	July–August 1993	Preliminary survey (1989–1990)	December 1994	Preliminary survey (1989–1990)	May–June 1999
No. examined	456	397	1278	1555	1734	975
No. positive	317	11	563	4	880	8
Prevalence	69.5%	2.8%	44.0%	0.2%	50.7%	0.8
Reduction	0	96.0%	0	99.5%	0	98.4%

Table 10. Changes in intensity of S. mekongi infection in certain villages

Villages monitored	Infection intensity (geometric mean eggs/gram faeces)	
	Preliminary survey (1989–1990)	After treatment with praziquantel
Sène Lam	47.6 (1989)	—
Ban Dong	40.5 (1989)	0.028 (1994)
Sène Had Gnay	47.6 (1989)	0.052 (1994)
Sène Had Noy	45.5 (1989)	—
Xieng Vang	31.5 (1989)	—
Kang Khong	38.2 (1989)	—
Ban Na	136.8 (1989)	—
Hin Siu	43.6 (1989)	0.04 (1994)
Ban Houay	49.0 (1989)	0.026 (1994)
Sène Tay	57.8 (1990)	0.026 (1999)
Som Van Ok	72.1 (1990)	0.44 (1999)
Tha Kham	59.6 (1990)	—
Tha Mak Hep	56.1 (1990)	0.30 (1999)

Table 11. Comparative change in prevalence of Schistosoma mekongi and Opisthorchis viverrini infections after treatment with praziquantel

Villages	S. mekongi prevalence		O. viverrini prevalence	
monitored	Preliminary survey (1989–1990)	After treatment with praziquantel	Preliminary survey (1989–1990)	After treatment with praziquantel
Ban Houay	40.4% (1989)	0.8% (1994)	42.1% (1989)	50.8% (1994)
Hin Siu	15.3% (1989)	1.1% (1994)	63.7% (1989)	58.6% (1994)
Sène Tay	8.2% (1989)	0.0% (1994)	630% (1989)	40.5% (1994)
Som Van Ok	76.9% (1990)	0.0% (1994)	41.5% (1990)	62.5% (1994)
Tha Kham	77.1% (1990)	0.0% (1994)	71.2% (1990)	59.7% (1994)
Tha Mak Hep	94.0% (1990)	0.0% (1994)	63.6% (1990)	54.2% (1994)

The change in prevalence was mirrored by a reduction in the intensity of infection (Table 10), indicating a stable situation had been achieved. However, continuing vigilance is essential to guard against both a resurgence of reinfection and importation of cases from Cambodia.

It should also be noted that mass treatment with praziquantel did not reduce the prevalence of opisthorchiasis, probably because it was not linked to a health education campaign (Table 11). Moreover, the dose of praziquantel used for schistosomiasis (40 mg/kg) is probably insufficient to treat opisthorchiasis: a higher dose would probably give better results.

Community-managed control of soil-transmitted helminthiasis in the Philippines

Y.C.F. Lumampao[1]
Soil-Transmitted Helminthiasis Control Programme
Communicable Disease Control Service
Department of Health, Manila, Philippines

L.C. David
Internal Planning Service
Department of Health, Manila, Philippines

Introduction

The Philippines archipelago lies off the coast of south-east Asia and comprises 7100 islands; Luzon, Visayas, and Mindanao are the three main islands. The climate is tropical, with a mean annual temperature of 27°C, and the estimated population is 75 million.

Status of soil-transmitted helminthiasis control in the Philippines

Soil-transmitted helminthiasis (STH) has long been a public health problem in the Philippines, until recently addressed only through community initiatives and clinical control. Various data show the prevalence rate of the disease in the country to range from 50% to 90%.

Soil-transmitted helminthiasis was the only disease concern identified as a public health problem in the 1993 National Health Plan, yet the Plan set no definitive policy for its control. In 1999, the Communicable Disease Control Service created the Soil-Transmitted Helminthiasis Control Programme (STHCP) to implement efficient control through mass deworming. The project site for initial activities was a province with a population of about one million people. At least five new provincial project sites are to be covered each year and, since the Philippines has 78 provinces, it will take at least 16 years to cover the entire country. The project duration is three years per provincial project site and will be conducted with the approval of local government units to ensure sustainability.

[1] Dr Y.C.F. Lumampao, Soil-transmitted Helminth Control Programme, Center for Infectious Diseases, 3rd Floor, Building 13, Department of Health, San Lazaro Compound, Santa Cruz, Manila 1003, Philippines
Tel/fax: +63 2 711 6808, E-mail: yvon@hotmail.com

Goal of the Control Programme

The goal of STHCP is to reduce morbidity, mortality, and other effects of soil-transmitted helminthiasis in the families and communities of each provincial project site. In general terms, the objectives of STHCP are to reduce the intensity of infection from high/medium to low in 100% of 2–14-year-old children in the project site within 3 years, and to reduce the prevalence of infection to <30% in the project site within 3 years.

The short-term objective is to treat 80% of children aged 2–14 years in the project site for a period of 3 years. Long-term objectives, which seek to sustain the reduction in infection intensity and in STH prevalence rates in project sites, include:

- ensuring the practice of good personal and food hygiene in 90% of households;
- increasing the number of sanitary toilets by 10% within 3 years; and
- promoting access to safe water in 80% of households.

Programme strategies

Mass targeted treatment[1]

Treatment of 2–14-year-old children is the main strategy of the programme. This age group is the most at risk and has the highest prevalence and intensity of infection (WHO, 1999). A study by Cabrera & Cruz (1983) in Mindoro, Philippines, showed that mass treatment of this age group reduces both the prevalence of the disease and the intensity of infection—or worm burden—in the individual, as well as lowering the frequency of STH transmission in the community.

Frequency of treatment was decided on the basis of rate of reinfection. Treatment once a year has very little effect on the prevalence of STH since reinfection 6 months after treatment is almost 100%; twice-yearly mass treatment at least is therefore needed (Cabrera & Valeza, 1980). The prescribed duration of treatment is 3 years because eggs remain viable in the soil and are infective for up to 2 years (Kobayashi, 1980; Kobayashi, Katakura & Hamada, 1986). At the end of 3 years, treatment should become selective provided that prevalence and intensity of infection have declined.

The "service outlet" for treating children aged 2–5 years is the family: under the supervision of a barangay health worker, the mother will give the anthelminthic drug to the child. For children aged 6–14 years, the service outlet is the school, where teachers, under the supervision of a school nurse, administer the drugs.

A drug delivery scheme was also identified. The Department of Health (DoH) provides the first dose to the Provincial Health Office (PHO) of the project site. The fourth dose will be provided to the PHO by a pharmaceutical manufacturer,

[1] Referred to subsequently as "mass treatment".

through the DoH, and the local government unit must develop a scheme for providing the second, third, and fifth doses.

The recommended drugs for project use are albendazole (400 mg) and mebendazole (500 mg), since both are in a chewable tablet form that provides a single, standardized dose. Generally, single-dose preparations are simpler and more convenient for mass treatment because teachers and non-health workers have to assist in their administration.

Community-managed programme

The hallmarks of a community-managed programme are:

- — networking between stakeholders, which will be formalized and facilitated through core groups formed at various levels, and
- — commitment to benchmarks.

Core groups serve as a forum for stakeholders to interface and to consolidate stakeholder strengths for better coordination and management of a project. All stakeholders must fulfil the benchmarks agreed on before proceeding to the next phase or activity of the project.

Family empowerment

The term "family empowerment" implies the development of family core groups and the development of five basic competencies in the families. Family competencies are a combination of knowledge, attitude, and practice (KAP) competencies that enable families to participate in the control of STH in an informed, cooperative, and capable manner.

Community organization involves forming clusters of about 20–25 families, each cluster having a lead family on whom information, education, and communication (IEC) will be focused. The lead family will provide a KAP model that other families in the cluster can follow. Each cluster will be supervised and monitored by a barangay health worker.

Environmental and personal hygiene

Good environmental and personal hygiene reduce the risk of transmission of infection and are therefore important for the achievement of long-term objectives and the sustained reduction of STH prevalence. Essential elements include:

- — the promotion of personal hygiene as a family KAP competency, through key habits such as washing hands before eating and after using the toilet, consistently wearing shoes and slippers;
- — construction of toilets for households without sanitary toilets;
- — promotion of the use of safe water as part of the family KAP competency.

Individual case management

Case management of non-target individuals will be primarily the responsibility of local government units. Community members aged 15 years and above will be given individual treatment.

Phased implementation

Phased implementation is essential if the yearly cost is to be kept at a manageable level—the cost of drugs for mass treatment is enormous. The first phase is pilot implementation to provide guidance on adjusting the project plan for national implementation and to develop basic programme tools. The activities in this phase include:

— prevalence survey to obtain baseline data on STH;
— human resources development to train regional health personnel in technology and management of the project;
— development of family competency tools, and monitoring and evaluation schemes;
— development of advocacy/IEC materials.

The second phase is national implementation by region, involving at least five provincial project sites per year.

Integrated implementation

Control of STH is being integrated with other mass treatment programmes, particularly those for filariasis and schistosomiasis. This will lead to greater efficiency in the use of resources: covering three diseases whenever possible avoids the duplication of activities, maximizes available personnel time, and keeps field travel costs to a minimum.

Community-managed STH control in the Philippines

Pilot site: Dipaculao, Aurora province

It was decided that the programme should be piloted in a small area. Selection criteria (the most important being the willingness of local government units to share resource requirements) were established and the municipality of Dipaculao, Aurora province, was chosen. Dipaculao is situated 340 km north of Manila and has a type IV climate, which means that rainfall is more or less evenly distributed throughout the year. It consists of 762 households and a total population of 3866; the average number of persons per household is five; 67% of households have sanitary toilets, while 11% have no toilets at all. Parasitism is ranked as second among leading causes of morbidity.

A meeting with the governor of the province and the municipal mayor was arranged by the provincial health officer to secure their approval for initiation of the STH control project in the area. The strategies and activities involved in STH control were presented to other stakeholders (nongovernmental organizations, the Department of Education, and other concerned agencies) in another forum. After the orientation, the interested parties organized a Provincial STH Control Committee (PSTHCC) and a Municipal STH Control Committee (MSTHCC). Family clusters were organized by the barangay health officials. The supervisor of the Department of Education was named as a member of the PSTHCC and all teachers were given an orientation on the STH control programme. The Committee then identified the donors for successive mass treatment doses. The first dose of deworming drugs was provided by the DoH; because of the acceptability of the control programme, the provincial government decided to provide the second dose and the municipal government the third. The fourth dose was financed by a pharmaceutical company; subsequently, the company undertook to provide the fourth dose for all future STH control projects in the country. The fifth dose was financed by each family once parents realized the significance of having worm-free children. World Vision, a nongovernmental organization with a number of other projects in the pilot area at the time, signalled its intention to provide the sixth dose.

A memorandum of agreement outlining the commitments made was signed by the Department of Health, the Department of Education, World Vision, the local government of Aurora province, and the pharmaceutical company.

A programme of training for the provincial and municipal health workers was undertaken. Health educators were taught how to develop an IEC plan and materials to build family competencies, providing families with the information necessary for them to make decisions regarding STH control and the health of their children. A KAP survey designed to establish baseline data was conducted in 400 households in the pilot area by volunteer health workers.

The provincial government organized a prevalence survey with the assistance of DoH personnel. Eight classrooms, each with an average of 40 pupils, were randomly chosen for stool examination. Pupils were instructed to bring stool samples, which were examined by a team of five provincial medical technologists under the supervision of a medical technologist from the DoH who trained them in the Kato–Katz technique.

Of 194 stool specimens, 125 (64%) were positive for the three common intestinal helminths (Table 1). The prevalence rate of *Trichuris trichiura* was highest, with 101 specimens positive (52%); *Ascaris lumbricoides* affected 95 (49%) children. This study also validated the finding of previous investigators that *Ascaris* and *Trichuris* are the commonest intestinal helminths in the Philippines. Hookworm infection was very rare in the study area, infecting only two children and categorized as light-intensity infection (Table 2). Overall, STH was slightly more prevalent in males (52%) than in females (48%).

Using the WHO guidelines for STH control (Montresor et al., 1998), the municipality of Dipaculao was classified as a Category 2 area, with a cumulative

Table 1. Prevalence rate of individual helminth infections in Dipaculao, Aurora province, 1998

Helminth	No. of positive specimens (*n* = 194)	Prevalence rate
Ascaris lumbricoides	95	49%
Trichuris trichiura	101	52%
Hookworm	2	1%
Cumulative prevalence rate	125	64%

Table 2. Intensity of infection of helminth infections in Dipaculao, Aurora province, 1998

Helminth	Infection intensity		
	Light	Moderate	Heavy
Ascaris lumbricoides	55 (58%)	36 (38%)	4 (4%)
Trichuris trichiura	88 (87%)	12 (12%)	1 (1%)
Hookworm	2 (100%)	—	—

prevalence of 64%. The recommended approach to STH control in a Category 2 areas is selective/targeted treatment of 2–14-year-old children, three times a year for a total of 3 years.

All children aged 2–14 years—of whom there were 2100 in the community—were given mass treatment; the first dose was provided by the DoH and the succeeding doses came from the sources identified in the memorandum of agreement. A repeat stool examination was carried out after the fourth dose.

Experience in STH control in Dipaculao led to the development of a *Handbook on the control of STH*, designed to address the basic needs of local health managers wanting to reduce STH prevalence in their communities. A companion volume—*Trainer's guide on the control of STH*—has also been developed to facilitate the training of local health workers. Thus, the STH control project in Dipaculao has not only helped the local people but has also benefited the national STH control programme.

Other activities

Other activities carried out during 1999—the start-up year for the STH control programme in the Philippines—included the following:

- Development of a *Reference manual on STH*, which will serve as a convenient reference for regional and provincial health workers who want to initiate STH control programmes.

- Proficiency training in parasite identification for medical technologists in 16 regions of the country.
- Training for all coordinators, health educators, and technical staff of the 16 regions, to inform them of the magnitude of the STH burden and provide them with an overview of the STH control programme.
- Prevalence and KAP surveys in another pilot area—San Fernando, Bukidnon—to test the feasibility of using the community pharmacy of the Family Health Management by and for Urban Poor Settlers (FAMUS) Programme as one of the service outlets in the dispensing of anthelminthic drugs. It is hoped that this will facilitate the participation of families in providing deworming drugs for their own children.

Activities in 2000

For the year 2000, scheduled activities included:

— expansion of pilot area in Dipaculao to cover the whole of Aurora province;
— implementation of STH control in five provinces;
— human resources development in the form of academic and field programmes for national experts; training in basic laboratory skills for STH control; and health information and community mobilization techniques.

The strength of the control programme in the Philippines lies in using simple STH control measures to expedite community-managed health programmes. The steps have been standardized to facilitate the implementation of STH control programmes by willing and dynamic communities. However, the need for different STH control strategies for more impoverished communities is recognized as an issue that has to be addressed.

Much remains to be done to convince people in national and local policy circles that investment in STH control is worthwhile. International health organizations may be of most help in stepping up advocacy in this area, emphasizing that:

— STH infections are the cause of the high disease burden among children 5–14 years old;
— mass treatment for STH is counted among the most cost-effective treatments for children;
— STH control, because of its immediate and visible effects, has been identified as a good entry-point programme and focus for integration in any community (World Bank, 1993).

References

Cabrera BD, Valeza F (1980) The reinfection rate of soil-transmitted helminths in the pilot areas after treatment. In: *Collected papers on the control of soil-transmitted helminthiases*, I:141–149. Tokyo, Asian Parasite Control Organization.

Cabrera BD, Cruz AC (1983) A comparative study on the effect of mass treatment of the entire community and selective treatment of children on the total prevalence of soil-transmitted helminthiases in two communities, Mindoro, Philippines. In: *Collected papers on the control of soil-transmitted helminthiases*, II:266–287. Tokyo, Asian Parasite Control Organization.

Kobayashi A (1980) Theory and practice applied in Japan for the eradication of *Ascaris* infection. In: *Collected papers on the control of soil-transmitted helminthiases*, I:233–242. Tokyo, Asian Parasite Control Organization.

Kobayashi A, Katakura K, Hamada A (1986) The fate of *Ascaris* eggs applied to the soil under various conditions. In: *Collected papers on the control of soil-transmitted helminthiases*, III:15–19. Tokyo, Asian Parasite Control Organization.

Montresor A et al. (1998) *Guidelines for the evaluation of soil-transmitted helminthiasis and schistosomiasis at community level.* Geneva, World Health Organization (document WHO/CTD/SIP/98.2).

WHO (1987) *Prevention and control of intestinal parasitic infections. Report of a WHO Expert Committee.* Geneva, World Health Organization, 1987 (WHO Technical Report Series, No. 749).

WHO (1999) *Report of the Informal Consultation on Monitoring of Drug Efficacy in the Control of Schistosomiasis and Intestinal Nematodes.* Geneva, World Health Organization (document WHO/CDS/SIP/99.1).

Experience and progress in controlling disease due to helminth infections in the Republic of Korea

Han-Jong Rim[1]
Korea Association of Health
Seoul, Republic of Korea

Introduction

Intestinal helminths are one of the most common causes of infections in humans, especially in tropical and subtropical countries. More than one billion of the world's population, including at least 400 million school-age children, are chronically infected with soil-transmitted helminths such as *Ascaris lumbricoides*, *Trichuris trichiura*, and the hookworms *Ancylostoma duodenale* and *Necator americanus* (WHO, 1998). These parasitic infections are more common in rural areas in the developing countries of Asia, Africa, and Central America, and are often linked to poverty and other social problems.

In the Republic of Korea, various helminthiases have been recognized as important endemic diseases for many years. In the past, the traditional application of night soil to vegetable gardens resulted in very widespread parasitism with a variety of helminths. In addition, eating habits that involve the consumption of raw vegetables, fish, crustaceans, and meat allow the transmission of helminth infections.

In the first nationwide survey of parasitic infections among Koreans, randomly selected stool examinations were conducted by the United States Army (Hunter et al., 1949). In the early 1950s, health problems were greatly aggravated by the Korean War. Public health officials did much to draw the attention of the Korean people to the prevention of both contagious diseases and parasitic infections. Shortly after the War, several reports were published on the prevalence of helminths among the population in different parts of the Republic of Korea (Brooke et al., 1956; Seo et al., 1969). It was widely recognized that clonorchiasis and paragonimiasis were important helminthic diseases in the country, in addition to ascariasis, hookworm infections, and trichuriasis.

The commonest and most widespread parasites were generally agreed to be the soil-transmitted helminths. Of these, *A. lumbricoides* was regarded as one of the

[1] Dr Han-Jong Rim, President, Korea Association of Health, 1097 Hwagok-Dong, Kangseo-Gu, Seoul 157-704, Republic of Korea
Tel: +82 2 2601 2016; Fax: +82 2 690 4905; E-mail: hjrim@kah.org.kr

most important because of its high prevalence and the risk of adverse nutritional and clinical effects of infection: prevalence among the population of the Republic of Korea was reported to be over 80% until the 1950s. Shin (1959), for example, reported the admission of 70 ascariasis cases requiring surgery for abdominal obstruction to a single local hospital in Seoul over a 5-year period during the 1950s.

It thus became apparent that there was an urgent need for nationwide efforts to control this infection and reduce its prevalence. Control activities against *A. lumbricoides* and other soil-transmitted helminths were carried out by the Korea Association for Parasite Eradication (KAPE—the forerunner of the present Korea Association of Health) under the law on prevention of parasitic diseases. All students and school pupils—who numbered more than 8 million and made up about one-fifth of the country's population—were subjected to twice-yearly faecal examinations. Individuals who were egg-positive for *Ascaris* were treated with anthelminthics such as piperazine, pyrantel pamoate, and mebendazole. Ascariasis control activities were continually stepped up, particularly from 1969 onwards, and were continued—with excellent results—until 1995.

The campaign targeted mainly pupils in primary, middle, and high schools and resulted in dramatic improvements in soil-transmitted helminthiasis. Stool examinations carried out between 1969 and 1995 revealed a remarkable decline in the prevalence of *A. lumbricoides* and other intestinal helminths: the infection rate of *A. lumbricoides* in students fell from 55.4% in 1969 (KAPE, 1969) to 0.02% in 1995 (KAH, 1995). The annual reduction rate was 1.8% up to 1973 but 4.7% thereafter, indicating that control efforts became more effective after 1973.

This paper describes the successful control of ascariasis and other helminthic diseases in the Republic of Korea, with special reference to control strategies and to progress in parasitic diseases control that has characterized the past three decades.

Approaches and progress in helminth control programmes

Organizations for control

The Korea Association for Parasite Eradication was founded in 1964 with strong support from the government and the clear objective of controlling parasitic infections. Its activities were carried out under the auspices of the Ministry of Health and Social Affairs. The headquarters in Seoul administered offices and laboratories in nine provinces and in two cities (Seoul and Pusan), supervising their control programmes for parasitic infections and maintaining close links with the Ministries of Health and Education and with scientific bodies such as the Korean Society of Parasitology and parasitology departments of university medical schools.

Legislation for the control of parasitic diseases

The Law for the Prevention of Parasitic Diseases was enacted in the National Assembly of the Republic of Korea in April 1966 and has formed the legal basis for the establishment of parasite control activities in the country. In the Article that deals with diagnosis and treatment of parasitic diseases, it is stipulated that schoolteachers or school principals are responsible for finding cases of parasitic infection—especially ascariasis—among pupils through stool examination and for treating them twice a year. A second important Article concerns the use and disposal of night soil, which is a source of significant problems for the control of soil-transmitted helminths. It has been recommended that night soil be treated in compost pits before being used, particularly in areas where it has traditionally been used to fertilize vegetable crops. Control of night soil use has been greatly facilitated by the rapid increase in the use of chemicals as alternative means of fertilizing crops.

Expert Committee for Parasite Control

In order to assess the problem, develop a proper plan, design a campaign, enforce regulations, and evaluate the results, the Ministry of Health and Social Affairs appointed an Expert Committee for Parasite Control in accordance with the 1966 Law for the Prevention of Parasitic Diseases. The Committee members are experts drawn from various fields, including the Korean Society for Parasitology, KAPE, and the Ministries of Education, Agriculture and Forestry, Internal Affairs, and Culture and Information.

Financial resources

The financial resources of KAPE (and now of KAH) are made up of membership dues, donations, and subsidies from central or local government. However, the amount derived from these is too meagre to finance all control activities. KAPE is therefore obliged to rely mainly on the fees for school-based faecal examinations. Under the law, the Association is authorized to perform these examinations twice a year. An individual fee of US$ 0.10 for stool examination was collected from each student by the Education Board of each school (with the permission of the mayor). No additional charge was made for treatment.

Technical personnel and facilities

The professional training of medical technologists is carried out at several junior colleges attached to the university medical colleges. After 2-year courses, the trainees take a national qualification test offered by the Ministry of Health and Social Affairs and, on passing, they are licensed to practise. Each provincial and city branch of KAPE should employ a certain number of licensed clinical technologists.

The space, materials, and equipment essential for maintaining adequate laboratory services must be available, together with facilities for communication and transport. The equipment for each laboratory is laid down by Ministry of Health regulations according to the size and locality of the laboratory. In areas where the population exceeds 300 000, the branch laboratory must be provided with at least eight microscopes, two incubators, and a refrigerator, should be at least 66 m^2 in area, and should be staffed by 14 technicians. Another technician should be added for every additional 100 000 population.

Foreign aids to KAPE

KAPE received assistance from the Overseas Technical Cooperation Agency (OTCA) of Japan for the purchase of vehicles, microscopes, other laboratory equipment, and anthelminthics over the period 1968–1974. Technical training programmes for laboratory technicians and management officers have also been conducted with the close cooperation of the Japan Association of Parasite Control under the sponsorship of OTCA.

Mass diagnosis and mass treatment

All stool specimens were examined by the cellophane thick-smear technique, known to be efficient, economical, and time-saving. The primary target of KAPE's mass treatment approach is ascariasis, plus control of hookworm infection on a lesser scale. Mass diagnosis therefore aims chiefly to detect egg-positive cases of ascariasis through the microscopic examination of faecal specimens collected twice a year (in spring and autumn) from pupils in primary, middle and high schools all over the country. A total of 12–16 million specimens were examined by annually by KAPE branch offices.

Until the early 1970s, ascariasis was treated with a santonin–kainic acid complex, together with piperazine derivatives. Mebendazole, synthesized by a local pharmaceutical company, was later used for mass treatment of both *Ascaris* and *Enterobius* infections; it proved effective even at a reduced single dose of 100 mg and has been widely and cost-effectively used in mass control programmes for ascariasis and enterobiasis.

Role of the local pharmaceutical company in parasite control

A local pharmaceutical company, established in 1962, has produced various pharmaceuticals using technology it has developed independently. It synthesized the raw material for mebendazole in 1975 and also successfully synthesized praziquantel in 1981. These drugs are the broad-spectrum anthelminthic agents for the treatment of intestinal nematodes as well as trematode and cestode infections. Using a totally different method of manufacture from that of the original makers of these drugs, the company was able to lower production costs. Its efforts have resulted in a domestic source of safe, effective, and low-cost medication. These factors were essential for the sustainability of the deworming campaign for more than 25 years.

Health education for parasite control

Through the mass media, KAPE provided the public with extensive health education about the importance of anthelminthic drug therapy and improved sanitation. Radio and television broadcasts have been devoted to the topic, and KAPE has made leaflets, posters, slides, and film strips available through schools, health agencies, etc. Liver, lung, and intestinal flukes—or trematodes—are acquired by eating raw, freshly pickled, or inadequately cooked freshwater fish, crayfish, or crabs containing metacercariae. The most practical method of avoiding these infections is to avoid eating the foods that carry them but, in the face of age-old traditions, it is extremely difficult to persuade people of this. Educational efforts should therefore be directed primarily towards schoolchildren, who are less entrenched in their food habits.

Evaluation of programme

In evaluating a control programme, it is difficult to measure progress in terms of improvements in the community's health and living standards. The results achieved by KAPE over the past two decades have therefore been measured in terms of prevalence rates over time. Figure 1 shows a pattern of declining ascariasis prevalence measured as percentages of egg-positive pupils in primary, middle, and high schools over the period 1969–1995. It can be concluded that the pupil-oriented ascariasis control project in Korea has been both very effective and efficiently carried out.

Figure 1. Declining pattern of *Ascaris* egg prevalence among pupils of primary, middle, and high schools, 1969–1996

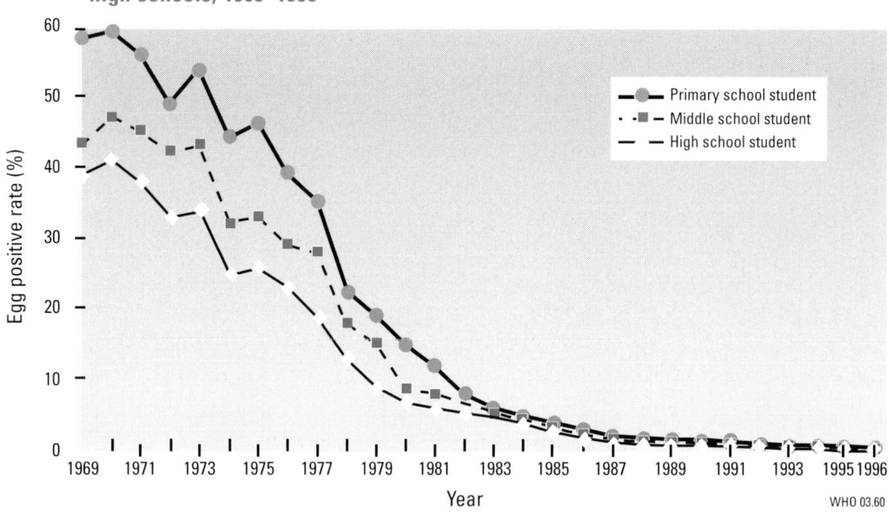

Figure 2. Changing pattern of parasite infections in Republic of Korea, 1971–1997

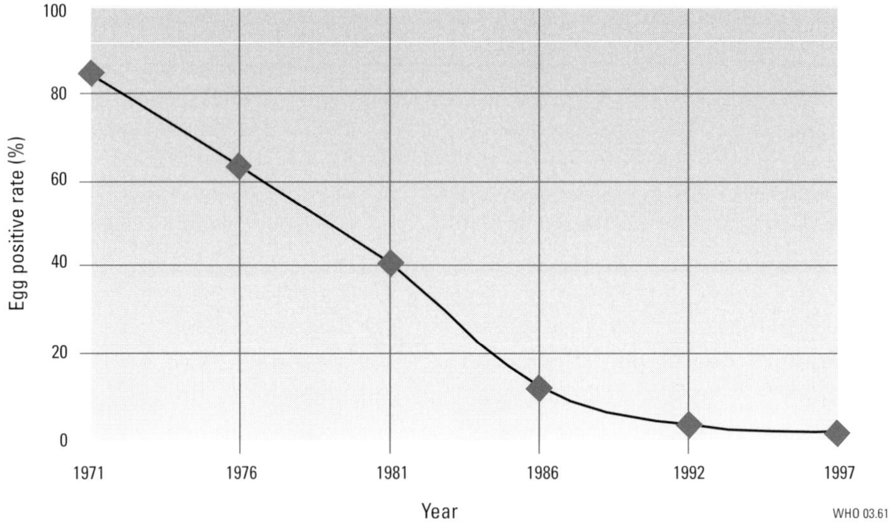

To evaluate the exact infection status and the standardized indicator of prevalence, a nationwide prevalence survey has been undertaken every 5 years: surveys of randomly selected sample populations in rural and urban areas were conducted by KAPE in 1971, 1976, 1981, 1986, 1992, and 1997. Results of these surveys reveal a steady and impressive decline in the prevalence of intestinal worm infections as a result of the control measures, from 84.3% of the 24 887 people examined in 1971 to 2.4% among the 45 832 examined in 1997 (Figure 2). Prevalence rates in rural populations are usually higher than those in urban populations.

Changing patterns of parasitic infections after the control measures

Soil-transmitted nematodes

As shown in Table 1, the egg-positive rate for *Ascaris lumbricoides* in people of all age groups nationwide was estimated by KAPE to be 54.9% in 1971, 41.0% in 1976, 13.0% in 1981, 2.1% in 1986, 0.3% in 1992, and 0.06% in 1997. Hookworm infection status has improved remarkably since the 1970s: the egg-positive rate (mostly *Ancylostoma duodenale*), was 10.7% in 1971, falling to 2.2% in 1976, 0.5% in 1981, 0.1% in 1986, 0.01% in 1992, and 0.007% in 1997. The prevalence of *Trichostrongylus orientalis* has followed a similar pattern to hookworm infection, with egg-positive rates of 7.7% in 1971, 1.0% in 1976, 0.2% in 1981, 0.02% in 1986, 0.004% in 1992, and zero in 1997.

It is thus clear that the overall prevalence of soil-transmitted helminthiases has declined significantly over the past two decades as a result of national control activities—repeated mass chemotherapy, plus environmental sanitation, health education, and growing production and use of chemical fertilizers in place of night soil. In addition, the rapid movement of the working population into industry,

Table 1. Prevalence of intestinal helminths in Republic of Korea, 1971–1997[a]

Helminth infection	1971	1976	1981	1986	1992	1997
Egg-positive rate	84.3	63.2	41.2	12.9	3.8	2.4
Ascaris lumbricoides	54.9	41.0	13.0	2.1	0.3	0.06
Trichuris trichiura	65.4	42.0	23.4	4.8	0.2	0.04
Hookworm	10.7	2.2	0.5	0.1	0.01	0.007
Trichostronglylus orientalis	7.7	1.0	0.2	0.02	0.004	—
Clonorchis sinensis	4.6	1.8	2.6	2.7	2.2	1.4
Metagonimus yokogawai	—	—	1.2	1.0	0.3	0.3
Taenia spp.	1.9	0.7	1.1	0.27	0.06	0.02
Hymenolepis nana	—	0.57	0.43	0.22	0.01	0.02

[a] Based on 5-yearly national surveys of intestinal helminth infections among the people of the Republic of Korea (Ministry of Health and Social Affairs, and Korea Association of Health, 1997).

the introduction of new agricultural technologies, the energetic reconstruction of rural villages, and enhanced educational opportunities are all thought to have contributed to improved socioeconomic conditions for the people of the Republic of Korea. However, the prevalence of intestinal parasite infections among inhabitants of remote islands remains high, making it important to strengthen control activities in efforts to achieve the final goal of "elimination".

Foodborne trematode infections

Foodborne trematode infections are acquired by eating raw or inadequately cooked fish or crustaceans containing parasite larvae. Three medically important species of trematodes—*Clonorchis sinensis*, *Paragonimus westermani*, and *Metagonimus yokogawai*—have long been known to be endemic in various areas throughout the Republic of Korea. Infection with these organisms is closely linked to human behavioural patterns, socioeconomic and cultural conditions, and the local biology and physical environment. According to the nationwide survey results, the prevalence of *Clonorchis sinensis* infection was 4.6% in 1971, 1.8% in 1976, 6% in 1981, 2.7% in 1986, 2.1% in 1992, and 1.4% in 1997. That is, prevalence was essentially unchanged over 20 years despite both the development of effective anthelminthics (Rim, 1986) and the implementation in 1984 of a government-funded, nationwide programme for the control of clonorchiasis using praziquantel. Widespread health education and further mass chemotherapy are needed if transmission of these trematode infections is to be significantly reduced.

Over the period 1984–1990, faecal samples from 3 009 166 people living in endemic areas were examined microscopically by the cellophane thick-smear method. Each egg-positive case of clonorchiasis was treated with a single dose of praziquantel (40 mg/kg body weight) at local health centres under the supervision of a physician (MHSA/KAH, 1992). The egg-positive rates were as follows:

13.3% of 168 877 people examined in 1984, 7.0% of 447 237 in 1985, 2.2% of 496 835 in 1986, 1.8% of 502 026 in 1987, 1.2% of 488 553 in 1988, 0.9% of 496 361 in 1989, and 0.9% of 409 277 in 1990. Not only were infection rates reduced in previously endemic areas, but also the proportion of heavy-intensity infection was lowered from 11.9% to 3.6%.

In addition to the mass treatment of risk groups, praziquantel was made available to all infected individuals. Other factors that seem to have contributed to blocking the reinfection cycle in many previously endemic areas include growing mechanization of agriculture, urbanization, industrialization, and water pollution. In addition, freshwater fish populations have declined because of the use of chemical fertilizers and pesticides, and the attitudes of local people towards eating raw fish have changed, although many people in certain river basin areas continue the practice.

Diagnosis of *Paragonimus westermani* infection is difficult, with the result that few data on its nationwide prevalence have been available. Prevalence is believed to have been quite high in the past and to have decreased significantly in recent years.

For *Metagonimus yokogawai*, the nationwide egg-positive rate has shown a declining tendency, from 1.3% in 1981 to 1.0% in 1986, 0.3% in 1992, and 0.3% in 1997. The prevalence of other intestinal trematodiases, such as heterophyiasis (*Heterophyes nocens*) and echinostomiasis (*Echinostoma hortense*, *E. cinetorchis*, and *Echinochasmus japonicus*), has remained steady or may be increasing. Infections with, for example, *Pygidiopsis summa*, *Heterophyopsis continua*, *Stellantchasmus falcatus*, *Centrocestus armatus*, *Stictodora fus*catus, and *Neodiplostomum seoulensis* have been detected in considerable numbers by identification of the adult worms (Rim, 1998). A new intestinal fluke, *Gymnophalloides seoi*, acquired by eating oysters, has been found to be prevalent in some south-western coastal areas (Lee, Chai & Lee, 1994), but only five cases of biliary or ectopic infections with *Fasciola hepatica* have been reported.

Foodborne cestode infections

Human taeniasis caused by *Taenia saginata* and *T. solium* is very common. The nationwide egg-positive rate for *Taenia* spp. (*T. solium* and *T. saginata* or *T. asiatica*), determined by faecal examination, was 1.9% in 1971, 0.7% in 1976, 1% in 1981, 0.3% in 1986, 0.006% in 1992, and 0.02% in 1997. A total of 29 worm-confirmed cases of *Diphyllobothrium latum* infection have been reported in the literature (Lee et al., 1983, 1989; Min, 1990) since the first case report of 1971 (Cho, Seo & Ahn, 1971). A case of *D. yonagoense* infection was also proved by recovery of an adult worm (Lee et al., 1988). Two extremely rare cases of intestinal infection with the adult worm of *Spirometra erinacei* have been reported (Lee et al., 1984), as have two cases of human infection with *Mesocestoides lineatus* (Eom, Kim & Rim, 1992). Of the larval cestode infections, more than 1000 clinical cases of cysticercosis have been reported in the literature, and many more cases go unreported. Hundreds of clinical cases of sparganosis have been reported, but again there may be many more cases that remain unreported (Min, 1990).

Other parasitic diseases

In the 1970s, a number of researchers reported the prevalence of *Enterobius vermicularis* infection, based on a single positive anal swab, to be in excess of 40% in the Republic of Korea. Even in the 1980s, the egg-positive rate in children remained at 30–40%. For *E. vermicularis*, nationwide surveys in all age groups revealed egg-positive rates of 12.0% in 1981, 3.6% in 1986, 0.9% in 1992, and 0.6% in 1997. However, the infection is thought to have decreased little compared with data obtained before the 1980s. Based on the single anal swab technique, it is generally agreed that the positive rate is approximately 2–10% in urban areas and 5–20% in rural areas, making *E. vermicularis* infection the most prevalent helminthic disease in the country now. The rates are highest in children aged 1–10 years.

The prevalence of *Hymenolepis nana* infection was 0.6% in 1976, 0.4% in 1981, 0.2% in 1986, 0.01% in 1992, and 0.02% in 1997; these figures show that this infection, too, has declined appreciably over the two decades. An early report on *Strongyloides stercoralis* cited an infection rate of 0.8%; there has been no subsequent case report, but several cases of hyperinfection syndrome with *S. stercoralis* were recorded recently in patients who had received long-term treatment with steroids. Cases of zoonotic parasitoses, such as hepatic and gastrointestinal capillariasis, have also been reported (Chae et al., 1993; Lee et al., 1993).

After the country's first reported case of *Anisakis* larval infection (of the tonsil), Im et al. (1990, 1995) recorded 241 cases of gastrointestinal anisakiasis. The infection is being increasingly reported nowadays.

Conclusion

Over the period 1969–1995, the Korea Association for Parasite Eradication conducted a parasite control programme that focused on the mass treatment of schoolchildren throughout the Republic of Korea. Repeated countrywide surveys have revealed the success of the programme in dramatically reducing the prevalence of ascariasis, trichuriasis, and hookworm infection. The programme has not significantly affected the prevalence of certain trematode infections; indeed, clonorchiasis remains an important public health problem, together with enterobiasis. It is acknowledged that, without appropriate legislation and overall government support, the programme would have a much lesser chance of success.

References

Brooke MM et al. (1956) Intestinal parasite survey of Korean Prisoner-of-War Camp. *US Armed Forces Medical Journal*, 7:708–714.

Chae GY et al. (1993) Hepatic capillariasis: first case report in the Republic of Korea. *American Journal of Tropical Medicine and Hygiene*, 48:610–625.

Cho SY, Seo BS, Ahn JH (1971) One case report of *Diphyllobothrium latum* infection in Korea. *Seoul Journal of Medicine*, 12:157–163.

Eom KS, Kim SH, Rim HJ (1992) Second case of human infection with *Mesocestoides lineatus* in Korea. *Korean Journal of Parasitology*, 30:147–150.

Hunter GW III et al. (1949) Parasitological studies in the Far East. VII: Epidemiological survey in southern Korea. *Journal of Parasitology*, 35(Suppl.):41.

Im KI et al. (1990) Gastric anisakiasis in Korea with review of 47 cases. *Yongsei Reports of Tropical Medicine*, 21:1–7.

Im KI et al. (1995) Gastric anisakiasis cases in Cheju-do, Korea. *Korean Journal of Parasitology*, 33:231–234.

KAPE (1969) *Results of the stool examination of the school students in Korea*. Seoul, Korea Association For Parasite Eradication (KAPE Monographic Series) (in Korean).

KAH (1995) *Results of the stool examination of the school students in Korea*. Seoul, Korea Association of Health (KAH Monographic Series):1–195 (in Korean).

Lee HS et al. (1983) First cases of *Dyphyllobothrium latum* infection. *Korean Journal of Parasitology*, 21:150–156 (in Korean).

Lee HS et al. (1993) A case of intestinal capillariasis in the Republic of Korea. *American Journal of Tropical Medicine and Hygiene*, 48:542–546.

Lee SH (1989) Seven cases of *Diphyllobothrium latum* infection. *Korean Journal of Parasitology*, 27:213–216 (in Korean).

Lee SH et al. (1984) Two cases of human infection by adults of *Spirometra erinacei. Korean Journal of Parasitology*, 22:66–71.

Lee SH et al. (1988) A case of *Diphyllobothrium yonagoense* infection. *Seoul Journal of Medicine*, 29:391–395.

Lee SH et al. (1994) High prevalence of *Gymnophalloides seoi* infection in a village of a south-western island of the Republic of Korea. *American Journal of Tropical Medicine and Hygiene*, 51:281–285.

MHSA/KAH (1992) *Evaluation of the mass treatment report of clonorchiasis in Korea (1983–1990)*. Seoul, Ministry of Health and Social Affairs/Korea Association of Heath (in Korean).

MHSA/KAPE (1971, 1976, 1981, 1986, 1992, 1996) *Prevalence of intestinal parasitic infections in Korea. First, second, third, fourth, fifth, sixth reports.* Seoul, Ministry of Health and Social Affairs/Korean Association for Parasite Eradication (KAPE Monographic Series) (in Korean).

Min DY (1990) Cestode infections in Korea. *Korea Journal of Parasitology*, 28(Suppl.):123–144.

Rim HJ (1986) The current pathobiology and chemotherapy of clonorchiasis. *Korean Journal of Parasitology*, 24(Suppl.):Monographic Series No. 3.

Rim HJ (1998) Field investigations on epidemiology and control of fish-borne parasites in Korea. *International Journal of Food Science and Technology*, 33:157–168.

Seo BS et al. (1969) Study on the status of helminthic infections in Korea. *Korean Journal of Parasitology*, 7:53–70 (in Korean).

Shin HK (1959) On surgical ascariasis. *Korean Journal of General Surgery*, 1:81–92 (in Korean).

WHO (1998) *It's a wormy world.* Geneva, World Health Organization (document WHO/CTD/SIP/98.4).

Control of soil-transmitted helminth infections in schoolchildren in Cambodia: implications for an integrated approach

C. Urbani, P. Odermatt,[1] D. Socheat
Médecins Sans Frontières, Phnom Penh, Cambodia

M. Sinoun
National Malaria Centre, Ministry of Health
Phnom Penh, Cambodia

S. Hoyer
World Health Organization, Phnom Penh, Cambodia

C. Hatz
Swiss Tropical Institute, Basle, Switzerland

Introduction

In recent years, academic institutions, medical agencies, and public health officers have shown an increasing interest in helminth infections, recognized as a priority health problem for large parts of the population in many developing countries. The 1993 *World development report* (World Bank, 1993) ranked intestinal helminths as one of the main causes of disease in children. WHO estimates that *Ascaris lumbricoides* and hookworms are among the 20 infectious diseases that kill the greatest number of people in the developing world: more than 200 million people are infected with schistosomiasis, around 1.4 billion by *A. lumbricoides*, 1.0 billion by *Trichuris trichiura*, and 1.3 billion by hookworms (Warren, Bundy & Anderson, 1990; WHO, 1996). These parasites are widely distributed in south-east Asia, and frequently occur in areas where lymphatic filariasis is transmitted.

High-transmission areas for *Schistosoma mekongi* have been known in Cambodia since 1993 (Biays et al., 1999), and the resulting severe disease was identified as a public health problem in need of immediate control measures. Médecins Sans Frontières (MSF)—already involved as the main partner in a rehabilitation pro-

[1] Dr P. Odermatt, L'Institut de la Francophonie pour la Médecine Tropicale, Vientiane, Lao People's Democratic Republic
Tel: +856 21 219346, Fax: +856 21 219347

gramme of provincial hospitals and health centres in endemic areas—has implemented a schistosomiasis control programme.

Surveys carried out during the schistosomiasis epidemiological assessment also aimed to define the prevalence and intensity of soil-transmitted infections and to confirm the transmission of lymphatic filariasis in the north-east of the country.

Epidemiology of helminth infections in Cambodia

Schistosoma mekongi infection is endemic in two Cambodian provinces, Kratie and Stung Treng, where the rocky banks of the Mekong River provide suitable conditions for the intermediate host (Goubert et al., 1994; Mouchet, 1995). Around 60 000 people are estimated to be at risk of infection, and in highly endemic areas the infection gives rise to serious disease (Biays et al., 1999; Odermatt P, Urbani C, unpublished reports). The most prevalent symptoms are related to portal hypertension (hepatosplenomegaly, ascites, oesophageal bleeding), but the infection also causes stunting, delay in sexual development, and—in the most advanced cases—wasting. In certain areas the severity of the problem required immediate control measures to address the high related morbidity and mortality.

The epidemiological assessment also revealed the extent of soil-transmitted helminth (STH) infections as a public health problem in Cambodia. Surveys carried out in 1997–1998 targeted schoolchildren, the group considered to be most affected; both urban and rural populations were investigated. Results are shown in Tables 1 and 2; Table 2 groups the results by intensity of infection.

Areas in which there is transmission of lymphatic filariasis were identified during surveys aimed at estimating the prevalence of schistosomiasis and other diseases (e.g. malaria), although the magnitude of the problem is not yet known. Epidemiological surveys are planned to assess the geographical distribution and the implications for the control and elimination.

The control programme

The approach taken to the problem in Cambodia was, first, to organize a framework for helminth control, followed by training at different levels for the health staff involved. Technical assistance and financial and logistic support in this preliminary phase were provided by Médecins Sans Frontières. Initial pilot activities achieved the setting up of a national control programme, with the National Malaria Centre (CNM) identified by the Ministry of Health as the programme's leading institution. The control programme was organized in accordance with a semi-horizontal model. National staff and selected provincial staff were involved in preparing a plan of action, defining main goals, objectives, strategies, and activities. Development, competencies, and implementation of control activities for helminth infection were highlighted in the programme goals. The major strategies were identified as periodic mass drug administration, training of health staff at central and peripheral level, and health education.

Table 1. Infection rates for soil-transmitted helminthiasis in schoolchildren in Cambodia

Village	*n*	*A. lumbricoides*		*T. trichiura*		Hookworms	
		Rate (%)	95% CI	Rate (%)	95% CI	(%)	95% CI
Achen	65	27.1	19.4–34.8	1.5	0–3.6	55.4	46.8–64.0
Benk Ket	69	55.1	46.7–63.5	2.9	0.1–5.7	24.6	17.3–31.9
Chatnaol	84	9.5	5.0–14.0	2.4	0.1–4.7	46.4	38.8–54.0
Float. Village	63	40.4	31.7–49.1	8.5	3.6–13.4	42.4	33.7–51.1
Hang Kosoun	74	45.9	37.8–54.0	5.4	1.7–9.1	21.6	14.9–28.3
Hang Savath	52	29.4	20.6–38.2	2	0–4.7	51	41.3–60.7
K. Chanh Tuk	48	70.8	61.6–80.0	2.1	0–5.0	66.7	57.2–76.2
Kamphan	51	60.8	51.2–70.4	3.9	0.1–7.7	52.9	43.1–62.7
Khampoun	114	32.5	26.4–38.6	14.9	10.2–19.6	50.9	44.3–57.5
Kenh Nhay	51	76.5	68.2–84.8	0	—	40.6	31.0–50.2
Koh Sneng	49	42.9	33.0–52.8	2	0–4.8	57.1	47.2–67.0
Lumphat	50	34	24.6–43.4	12	5.6–18.4	46	36.1–55.9
Maroeum	80	67.5	60.2–74.8	1.3	0–3.1	55	47.2–62.8
Nhang Sum T	95	43.2	36.1–50.3	6.3	2.8–9.8	69.5	62.9–76.1
Oray	48	58.3	48.3–68.3	4.2	0.1–8.3	66.7	57.2–76.2
Pha Bung	67	43.3	34.8–51.8	3	0.1–5.9	47.8	39.3–56.3
Pong Ro	80	71.3	64.2–78.4	22.5	16.0–29.0	57.5	49.8–65.2
Prek Chik	83	50.6	42.9–58.3	18.1	12.2–24.0	47	39.3–54.7
Sambok	142	16.2	11.9–20.5	2.1	0.4–3.8	45.1	39.3–50.9
Samdech Av	79	39.2	31.5–46.9	7.6	3.4–11.8	46.8	38.9–54.7
Siem Pang	105	47.6	40.8–54.4	3.8	1.2–6.4	43.8	37.0–50.6
Sre Khoeurn	58	13.8	7.5–20.1	3.4	0.1–6.7	50	40.8–59.2
Tabauk	30	3.3	−1.3–7.9	44.8	32.1–57.5	83.3	73.8–92.8
Taveng	69	24.6	17.3–31.9	0	—	78.3	71.4–85.2
Thmar Reap	80	42.5	34.8–50.2	3.8	0.8–6.8	48.8	41.0–56.6
Veal Ksach	43	41.9	31.4–52.4	9.3	3.1–15.5	37.2	26.9–47.5
Chram Chamres	345	64.3	60.7–67.9	30.4	26.9–33.9	14.8	12.1–17.5
Battambang	140	80	75.3–84.7	10.7	7.0–14.4	86.4	82.3–90.5
Siem Reap	135	69.6	64.1–75.1	16.2	11.8–20.6	75.5	70.3–80.7
Mossakrong	98	59.2	52.2–66.2	0	—	22.4	16.5–28.3
Total	*2547*	*45.4*	*44.0–46.8*	*8.2*	*7.4–8.9*	*51.1*	*49.7–52.4*

Table 2. Intestinal nematodes in 2547 schoolchildren in Cambodia, grouped by infection intensity

Infection intensity	*A. lumbricoides*		*T. trichiura*		Hookworms	
	%	95% CI	%	95% CI	%	95% CI
Light	63.2	62.4–64.1	91.2	90.5–91.8	85.5	85.0–86.1
Moderate	35.3	34.5–36.1	8.8	8.1–9.4	7.8	7.5–8.2
Heavy	1.3	1.2–1.5	0	—	6.5	6.2–6.8

The CNM is now charged with maintaining a network of provincial health departments, nongovernmental organizations (NGOs), and other interested bodies participating in the programme. The programme has a team manager, and a leader has been identified for each component (schistosomiasis, STH, filariasis). The team works in close cooperation with local government and community representatives, NGOs, and the school system.

The national referral institution, the CNM, represents the "core" of the control programme, and the management staff should improve awareness among policy-makers at the government level and strengthen cooperation with international agencies and donors. CNM should also be responsible for the management of drugs and health education materials and their distribution to provincial centres and to the partners. In addition, it will monitor and evaluate the project by collecting information on schistosomiasis and helminth infections and associated morbidity in the schoolchildren. The network coordinator organizes meetings at which methods of intervention and collaboration can be discussed by staff of provincial health and education departments and NGOs. Data collection, health education, and mass drug administration are the specific responsibilities of the peripheral health system. Control activities are planned and organized at this level, under the responsibility of the provincial team leader.

Training was organized at different levels. The first step, organized at the CNM, was the training of trainers and focused on control strategies and activities. Next, the training of provincial staff covered transmission, laboratory diagnosis, treatment strategies, and passive and active case-detection. Third, training of education sector staff aimed to provide them with a basic understanding of parasite infection and teach them how to prevent infection, conduct a questionnaire survey for schistosomiasis, and administer tablets.

General policy decisions were taken at the central level, but management of the programme was local. All the control activities were completely integrated in the activities of the peripheral health system. Provincial teams visiting villages to deliver drugs sometimes also undertook impregnation of bednets for malaria control: in Cambodia, integration of schistosomiasis and malaria control activities has reinforced the efficacy of each programme. This multi-intervention approach was particularly valuable for offering services to populations in remote areas.

Malaria control is the main objective for the CNM, with distribution of mosquito nets the main strategy; however, the infrastructure has also been invaluable for the delivery of other interventions, such as immunization, malaria treatment, helminth control, lymphatic filariasis survey, vitamin A distribution, leprosy screening, and iodine supplementation.

Integration permits optimal use of staff and finances, particularly where resources are very limited. Moreover, the principle of integrating disease control activities to save resources is easily understood by peripheral health workers. The

CNM furnishes the essential support for data analysis and the periodic updating of guidelines.

The national control programme is currently supported by WHO, which provides technical advice and facilitates contact with potential external donors.

Control strategies

A rapid assessment method was used to identify areas at risk of schistosomiasis transmission. School-age children were then targeted by a control strategy of periodic mass administration of a single dose of praziquantel (40 mg/kg body weight). A tool for determining the correct dose of praziquantel without the need for weighing scales was tested in an effort to improve the coverage of the programme and to extend drug distribution to remote areas. A "bench aid" was prepared, with pictures of children grouped according to their weight and with the correct number of tablets of praziquantel shown next to each picture (Figure 1). Two pilot schools were used in validating this tool, which proved to be highly reliable.

Drug distribution combined with health education has dramatically reduced infection rates—prevalence in four sentinel villages has fallen from an average of 70% to less than 10%—and provided excellent control of morbidity (Urbani & Odermatt, Médecins Sans Frontières internal report). Moreover, ultrasound surveys in endemic villages have shown a declining rate of severe liver fibrosis (Hatz & Urbani, unpublished data).

Good short-term results have been obtained in pilot areas with large-scale mass chemotherapy. Tables 3 and 4 show the results of twice-yearly treatment with a single 500-mg dose of mebendazole. As shown in Tables 3 and 4, twice-yearly treatment with a single 500-mg dose of mebendazole progressively reduced the intensity of infection, which is normally associated with a reduction in morbidity. In schistosomiasis-endemic areas, praziquantel was given in addition to the single dose of mebendazole.

Discussion

There is evidence that three main strategies are effective in reducing the burden of human helminth infections—periodic mass drug administration, strengthening of knowledge of the disease and relevant skills in the public health system at central and peripheral level, and health education and information campaigns (WHO, 1990; Albonico et al., 1995). Moreover, significant advantages in cost and sustainability can be gained from the integration of control programmes for all intestinal parasitic diseases, which have much in common with regard to epidemiology, diagnosis, community treatment, and prevention.

Extensive experience in tropical countries confirms that, for example, schistosomiasis control improves helminth control in general. Combined chemotherapy (e.g. praziquantel plus albendazole or mebendazole) is proposed for the control

Figure 1. Bench aid for providing praziquantel where there are no scales

Note: This is the English translation of the Khmer chart. On the back, pictures cover weights of up to 50 kg.

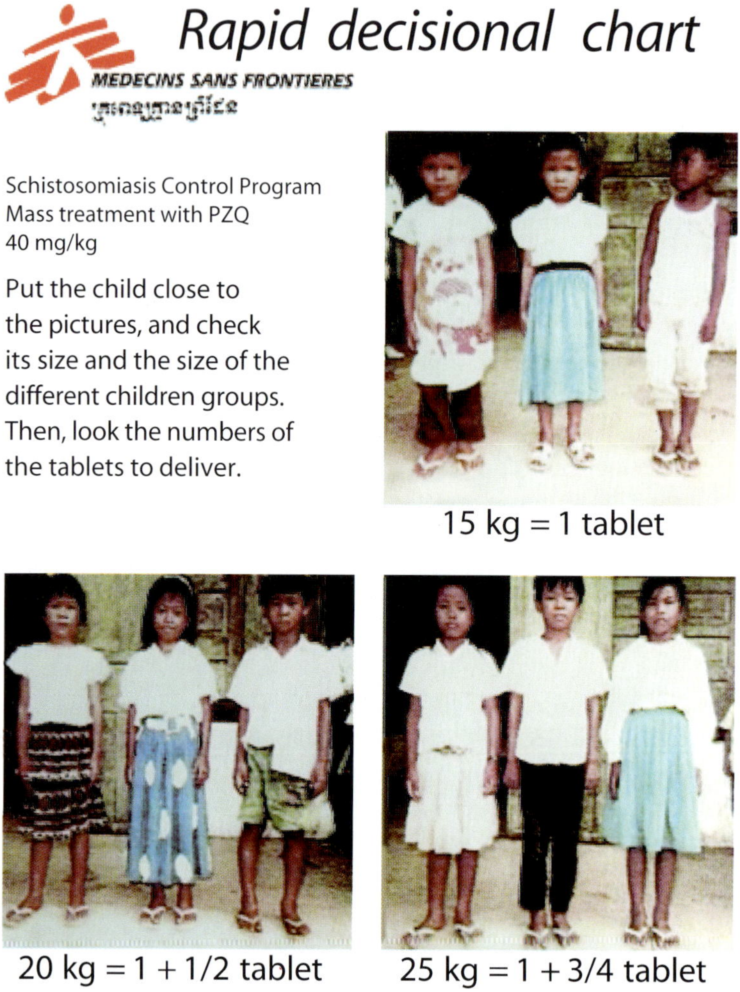

of mixed helminth infections (Adewunmi et al., 1993; Albonico et al., 1997; Magnussen et al., 1997), and the addition of ivermectin will also control filariasis. More interesting still is the integration of these programmes with sanitation or health education programmes, as well as nutritional, veterinary, and public health measures. However, integration must be combined with decentralization: the principal activities should be integrated into the peripheral health system. A peripheral health facility in charge of epidemiological surveillance should also be the focal point for the control measures, such as mass treatment campaigns and health education.

Table 3. Village of Sdau: prevalence and intensity of soil-transmitted helminth infections in school-age children following mebendazole 500 mg single-dose mass delivery every 6 months

Helminth	Baseline data ($n = 184$)		6 months ($n = 121$)		12 months ($n = 105$)		24 months ($n = 134$)	
	Prevalence	Intensity[a]	Prevalence	Intensity[a]	Prevalence	Intensity[a]	Prevalence	Intensity[a]
A. lumbricoides	37.2	3752.8	24.0	1631.4	22.9	859.8	33.7	401.0
T. trichiura	6.8	3.7	3.3	1.1	0	—	1.6	1.2
Hookworms	79.7	634.6	64.5	284.0	47.6	129.8	45.1	227.5
Total	*83.8*	—	*68.6*	—	*58.1*	—	*71.4*	—

[a] Arithmetic mean.

Table 4. Urban area of Chraing Chamres, Phnom Penh: prevalence and intensity of soil-transmitted helminth infections in school-age children following mebendazole 500 mg single-dose mass delivery every 6 months

Helminth	Baseline data ($n = 297$)		4 months ($n = 310$)		6 months ($n = 282$)		1224 months ($n = 441$)	
	Prevalence	Intensity[a]	Prevalence	Intensity[a]	Prevalence	Intensity[a]	Prevalence	Intensity[a]
A. lumbricoides	66.7	7021	21.3	187.2	27.3	1050.8	31.0	281.6
T. trichiura	32.4	102.8	5.3	3.4	4.9	24.6	7.4	9.5
Hookworms	16.2	64.9	6.0	1.4	7.4	1.2	4.1	12.5
Total	*88.3*		*30.3*		*35.8*		*39.6*	

[a] Arithmetic mean.

In the current initiative by WHO and the United Nations Children's Fund (UNICEF) to advance the control of STH in Cambodia, CNM sees its role as providing technical support for the evaluation, monitoring, and planning unit. A large pilot control programme at the provincial level is recognized as the next step towards a nationwide, school-based deworming campaign.

As high prevalence and intensity of STH infections are associated with risk of morbidity, mainly in school-aged children, treatment of this population group is the most feasible and cost-effective strategy (Hall et al., 1997). Recently, the use of the education sector for health delivery has been stressed as a way of improving the health of schoolchildren. In developing countries there are usually more schools than clinics—and more teachers than health workers. The cost-effectiveness of school health programmes as a public health strategy has been borne out by analysis (World Bank, 1993). Government investments in education in developing countries are compromised if children are absent from school or less able to learn as a result of ill health (The Partnership for Child Development, 1997). Provided that school attendance rates were quite good, school-based programmes in Cambodia would be effective in improving the helminth infection status of school-aged children. School-based programmes have also been reported to be effective both in reaching non-enrolled school-age children and in communicating health education messages to the whole population (Del Rosso & Marek, 1996; Montresor et al., 1998). Children can become "agents of change", serving as entry points to reach entire communities.

Training courses on the health impact of STH and their prevention will be established in collaboration with UNICEF and selected pilot schools. Health education materials that will facilitate teaching will also be developed: during 1999 there was good experience with the use of flip charts in the schistosomiasis control programme.

The intervention package for STH will include regular treatment with anthelminthic drugs (once or twice a year) and health education sessions. Every school—in pilot provinces initially—will be provided with a deworming kit. This is a box containing everything necessary for deworming, and providing health education to, 1000 children; it includes 1000 tablets of mebendazole (500 mg) and of vitamin A, forms, educational material, a health education book, and an educational game.

Helminth infections are a substantial challenge to health improvement in many developing countries and a significant obstacle to development in the poorest of those countries. Experience in Cambodia lends substance to the claim that there are sustainable solutions to the problem of helminth infections (Savioli, Bundy & Tomkins, 1992).

References

Adewunmi CO et al. (1993) Schistosomiasis and intestinal parasites in rural villages in southwest Nigeria: an indication for expanded programme on drug distribution and integrated control programme in Nigeria. *Journal of Tropical Medicine and Parasitology*, 44:177–180.

Albonico M et al. (1995) A randomized controlled trial comparing mebendazole and albendazole against *Ascaris*, *Trichuris* and hookworm infections. *Transactions of the Royal Society of Tropical Medicine and Hygiene*, 88:585–589.

Albonico M et al. (1997) Parasitic infections in Pemba Island school children. *East African Medical Journal*, 74:294–298.

Biays S et al. (1999) A focus of *Schistosomiasis mekongi* rediscovered in Northeast Cambodia: cultural perception of the illness; description and clinical observation of 20 severe cases. *Tropical Medicine and International Health*, 4:662–673 (in French).

Del Rosso JM, Marek T (1996) *"Class action": improving school performances in the developing world through better health and nutrition.* Washington, DC, World Bank (Directions in Development Series).

Goubert L et al. (1994) Foyer de *Schistosoma mekongi* dans la province de Kratie, Cambodge du nord. [Focus of *Schistosoma mekongi* in Kratie province, northern Cambodia.] *MSF Medical News*, 3:15–19.

Hall A et al. (1997) Promoting child health through helminth control—a way forward? *Parasitology Today*, 13:411–413.

Magnussen P et al. (1997) A school-based approach to the control of urinary schistosomiasis and intestinal helminth infections in children in Matuga, Kenya: impact of a two-year chemotherapy programme on prevalence and intensity of infections. *Journal of Tropical Medicine and International Health*, 2:825–831.

Montresor A et al. (1998) *Guidelines for the evaluation of soil-transmitted helminthiasis and schistosomiasis at community level.* Geneva, World Health Organization (document WHO/CTD/SIP/98.1).

Mouchet F (1995) Malacological data on the transmission of *Schistosoma mekongi* in Cambodia. In: *Proceedings of the European Conference on Tropical Medicine, Hamburg, October 1995.* Bonn, Deutsche Stiftung für Internationale Entwicklung.

Savioli L, Bundy DAP, Tomkins A (1992) Intestinal parasitic infections: a soluble public health problem. *Transactions of the Royal Society of Tropical Medicine and Hygiene*, 86:353–354.

The Partnership for Child Development (1997) Better health, nutrition and education for the school-aged child. *Transactions of the Royal Society of Tropical Medicine and Hygiene*, 91:1–2.

Warren KS, Bundy ADP, Anderson RM (1990) Helminth infections. In: *The World Bank health sector priorities review.* Washington, DC, World Bank: Chapter 15.

World Bank (1993) *World development report 1993: investing in health.* New York, Oxford University Press.

WHO (1990) *Informal Consultation on Intestinal Helminth Infections, Geneva, 9–12 July 1990.* Geneva, World Health Organization (document WHO/CDS/IPI/90.1).

WHO (1996) *World Health Report 1996. Fighting disease, fostering development.* Geneva, World Health Organization.

Effectiveness of deworming in schoolchildren through school feeding in Indonesia

V. Hadju,[1] A.R. Thaha
Centre for Food, Nutrition and Health
Research Centre and School of Public Health
Hasanuddin University, Makassar, Indonesia

Z. Agus
School of Medicine, Andalas University
Padang, Indonesia

F. Jalal
Centre for School Feeding Project
National Development Planning Agency
Jakarta, Indonesia

Introduction

The prevalence of soil-transmitted helminthiases is still high in most developing countries, and recent studies have revealed a high prevalence of *Ascaris lumbricoides* and *Trichuris trichiura* infections, especially in urban slum areas. Consistently, the highest prevalence and intensity of both *A. lumbricoides* and *T. trichiura* infections are to be found in school-age children (Bundy, Kan & Rose, 1988). Surveys in slum areas of Makassar, South Sulawesi, Indonesia, found that 92% and 98% of schoolchildren had *A. lumbricoides* and *T. trichiura* infections, respectively (Hadju et al., 1995). Corresponding figures for schoolchildren in rural areas of Sukaraja District in West Java were 76% and 44% (Pegelow et al., 1997).

The impact of *A. lumbricoides* and *T. trichiura* infections on health and nutritional status is substantial: growth, appetite, activity, and physical fitness improve appreciably following treatment (Latham et al., 1990; Stephenson et al., 1993; Adams et al., 1994; Hadju et al., 1996a). Several studies have also noted improved intellectual development and cognitive performance after deworming treatment of schoolchildren (Nokes et al., 1991, 1992; Simeon et al., 1994; Hadidjaya et al., 1996). Helminth infections have been shown to impair both plasma iron

[1] Dr Veni Hadju, Pusat Studi Gizi dan Pangan, dan Kesehatan, Gedung Pusat Kegiatan Penelitian, Lt. 2, Kampus Unhas Tamalanrea, Makassar 90245, Indonesia
Tel: +62 411 585705, Fax: +62 411 584386, E-mail: Micronut@TheOffice.net

concentrations (Karyadi et al., 1996) and the absorption of β-carotene (Jalal et al., 1998).

Targeted chemotherapy of schoolchildren has proved successful in controlling helminth infections (Bundy et al., 1990; Thein-Hlaing et al., 1990), and deworming of schoolchildren is recommended by WHO (1992a, 1992b). Indonesia currently operates a countrywide, twice-yearly deworming programme in conjunction with a school feeding programme. Large-scale deworming began in 1996/1997, initially covering all schoolchildren in poor areas outside Java and Bali provinces (16794 schools with 2100000 children). In 1999, the programme expanded to cover all provinces—60636 schools with 9809179 children. Three times a week, children are given a mid-morning snack containing approximately 5 g of protein and 300 $kcal_{th}$.[1] An anthelminthic drug is provided once every 6 months: albendazole and pyrantel are used alternately. All children are given the anthelminthic drug—no screening is done. The objective of this paper is to report the results of an evaluation designed to assess the impact of the anthelminthic treatments on the prevalence and severity of *A. lumbricoides* and *T. trichiura* infections within the context of the school feeding programme.

Material and methods

Study location, population, and design

The study evaluation was conducted in South Sulawesi and West Sumatra—provinces in the eastern and western parts of Indonesia, respectively. Two districts in each province were involved, representing rural and urban areas. In South Sulawesi, six primary schools (four covered by the programme and two not covered) were studied in each district; four primary schools (two covered by the programme and two not) were selected in each of the chosen districts in West Sumatra. Thus, a total of 12 primary schools in South Sulawesi and eight in West Sumatra were included in the study. They were selected according to their location (rural and urban areas or coastal and highland areas) and sanitary conditions (slum and periurban areas). A further criterion for selection was that schools should have participated in the programme for at least one year before the start of the study.

Children aged 6–11 years were selected from the schools: each school was represented by 100–120 children from grade II to grade V. Grade I children were excluded because they had never been covered by the programme, and grade VI children would leave school during the course of the study. Where the number of available children was greater than needed, individuals were systematically chosen from the school's records.

The study design was different in the two provinces. In West Sumatra, the study was cross-sectional; that in South Sulawesi was longitudinal.

[1] 1 $kcal_{th}$ = 4.18 kJ.

In the longitudinal study, data on prevalence and intensity of helminthiases were assessed five times—at baseline (exam 1), 3 weeks after the first dose (exam 2), 6 months before the second dose (exam 3), 3 weeks after the second dose (exam 4), and at 12 months (exam 5). Exams 2 and 4 were performed only for those who returned their stool samples at exams 1 and 3, respectively. Anthropometric data, absenteeism, and school records were examined on three occasions (at baseline, 6 months, and 12 months; data not shown). For this analysis only four schools (instead of six) were studied in each district.

Generic formulations of the anthelminthic drugs albendazole and pyrantel were administered to children in the morning on school days: a single dose of albendazole (400 mg) was given in the first semester and pyrantel pamoate (10 mg/kg body weight) in the second semester. Both drugs were produced by the state-owned pharmaceutical company. The alternate use of two different drugs in this programme was a decision of the committee of the School Feeding Project and was based on expert advice on avoidance of drug resistance.

Parents of all children were informed about the study and asked to consent to their children's participation. Permission from provincial and local government, local health institutions, and the principals and teachers in all selected schools were obtained in advance of the study. For ethical reasons, children in control areas were given a dose of pyrantel at the end of the study.

Data collection and analyses

Field workers distributed bottles labelled with identification numbers to the children, with instructions on how to use the bottles. Children were asked to return the bottles with stool samples the next morning. Samples were taken to the parasitology laboratory at the Medical School of Hasanuddin University, Makassar, where they were examined for parasite eggs by two trained technicians using the modified Kato–Katz technique recommended by WHO (1991).

Data analyses were performed using SPSS (Win 7.5) software. Within groups, changes in the prevalence and intensity of both *A. lumbricoides* and *T. trichiura* infections between two examinations were assessed by McNemar's test and a paired Student's *t*-test. Differences between groups were assessed by a chi-square test and an independent Student's *t*-test. The intensities of helminth infections, which have negative binomial distribution, were transformed by conversion to natural logarithms ($\log x + 1$) wherever possible. The percentage egg reduction rates of *A. lumbricoides* and *T. trichiura* infection between the two exams was calculated from geometric mean counts of eggs per gram (epg) using the formula: % egg reduction = [(initial epg − final epg)/initial epg] × 100%. Cure rate was calculated from the formula: cure rate (%) = [(initial prevalence − final prevalence)/initial prevalence] × 100%.

Changes in prevalence and infection intensity between exams 1 and 2 and exams 3 and 4 were assessed for the efficacy of albendazole and pyrantel, respectively. At the time of analysis, infection intensities of *A. lumbricoides* and *T. trichiura* were categorized as mild, moderate, and severe, according to WHO criteria (WHO, 2002).

Results

Baseline data

Overall, the prevalences of *A. lumbricoides* and *T. trichiura* infections were 51.2% and 70.4%, respectively, in South Sulawesi and 43.3% and 49.8% in West Sumatra. In South Sulawesi, the prevalence of *A. lumbricoides* in dewormed areas was lower than that in control areas (48% vs 58%, $\chi^2 = 13$, df = 1, $P = 0.0001$). However, there was no significant difference between dewormed and control areas for *T. trichiura* (71% vs 70%). In West Sumatra, by contrast, the prevalence of *A. lumbricoides* was significantly lower in dewormed than control areas (31% vs 57%, $\chi^2 = 56$, df = 1, $P = 0.0001$) while that of *T. trichiura* was significantly higher in dewormed than control areas (54% vs 45%, $\chi^2 = 6.8$, df = 1, $P = 0.009$). Figure 1 shows the prevalences of *A. lumbricoides* and *T. trichiura* in dewormed and control areas categorized by rural and urban locations. In both provinces, both

Figure 1. Prevalence of Ascaris *lumbricoides* (A) and *Trichuris trichiura* (B) at baseline in South Sulawesi and West Sumatra, categorized by location (dark green = rural, light green = urban)

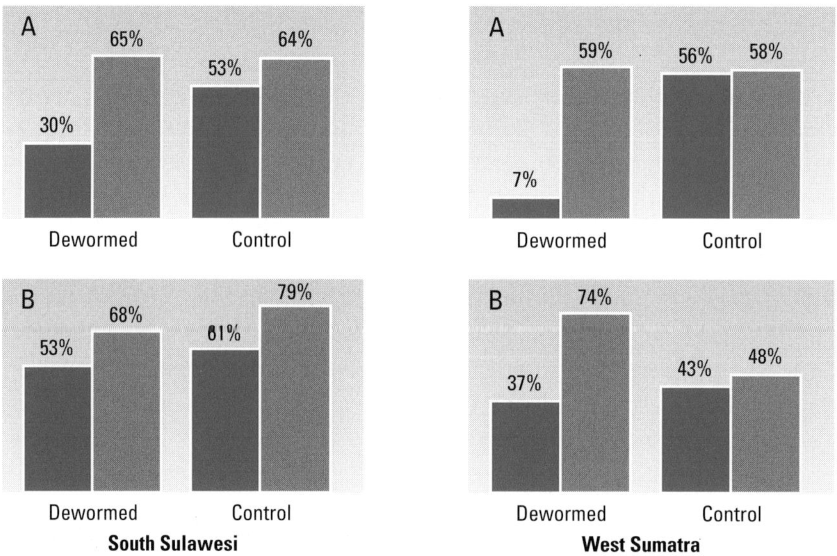

WHO 03.78

In South Sulawesi, there were significant differences between dewormed and control areas for *A. lumbricoides* ($\chi^2 = 13$, df = 1, $P = 0.0001$) but not for *T. trichiura* ($\chi^2 = 0.13$, df = 1, $P = 0.7$). In West Sumatra, however, there were significant differences between dewormed and control areas for both *A. lumbricoides* ($\chi^2 = 56$, df = 1, $P = 0.0001$) and *T. trichiura* ($\chi^2 = 7$, df = 1, $P = 0.009$).

Table 1. Frequency distribution of intensities of *Ascaris lumbricoides* and *Trichuris trichiura* in each study area at baseline

Area	*A. lumbricoides* infection intensity (%)[a]				*T. trichiura* infection intensity (%)[a]			
	None	Light	Mod.	Sev.	None	Light	Mod.	Sev.
South Sulawesi								
Dewormed areas								
Rural ($n = 442$)	70	24	4	1	47	53	1	0
Urban ($n = 444$)	35	37	21	7	12	68	9	12
Subtotal ($n = 886$)	52	31	13	4	29	60	5	6
Control areas								
Rural ($n = 220$)	47	40	11	2	40	59	1	1
Urban ($n = 217$)	36	31	17	16	21	60	12	8
Subtotal ($n = 437$)	42	36	14	9	30	59	6	4
West Sumatra								
Dewormed areas								
Rural ($n = 228$)	93	6	0	1	63	37	0	0
Urban ($n = 198$)	41	21	10	27	26	72	2	0
Subtotal ($n = 426$)	69	13	5	13	46	53	1	0
Control areas								
Rural ($n = 210$)	44	26	7	22	58	41	1	0
Urban ($n = 191$)	42	26	9	22	52	48	1	0
Subtotal ($n = 401$)	43	26	8	22	55	44	1	0

[a] Chi-square test between dewormed and control areas in South Sulawesi: *A. lumbricoides* ($\chi^2 = 19$, df = 3, $P = 0.0001$) and *T. trichiura* ($\chi^2 = 3.2$, df = 3, $P = 0.364$). In West Sumatra: *A. lumbricoides* ($\chi^2 = 57$, df = 3, $P = 0.0001$) and *T. trichiura* ($\chi^2 = 7$, df = 2, $P = 0.031$).

prevalence and the intensity of *A. lumbricoides* and *T. trichiura* infection are strikingly higher in urban than in rural schools.

Frequency distributions of infection intensity are shown in Table 1, which shows that percentages of moderate and severe infections of *A. lumbricoides* and *T. trichiura* were higher in urban than in rural areas in all groups. Differences of this sort detected at baseline must be taken into consideration in interpreting the results of the evaluation.

Efficacy of generic anthelminthic drugs

For evaluation of efficacy, faecal examinations were performed before and 3 weeks after drug administration. Four schools in each dewormed and control area were involved in the study. Table 2 shows the cure rate achieved with albendazole. At baseline, prevalence of *A. lumbricoides* and *T. trichiura* was different in each school. The highest prevalences of *A. lumbricoides* and *T. trichiura* were observed in School 1 (86% and 100%, respectively) and School 5 (87% and 97%, respectively); the lowest prevalences of *A. lumbricoides* (27%) and *T. trichiura* (51%)

Table 2. Cure rate in *Ascaris lumbricoides* and *Trichuris trichiura* infections 3 weeks after administration of albendazole

Primary school[a]	*A. lumbricoides*			*T. trichiura*		
	% positive		Cure rate (%)	% positive		Cure rate (%)
	Exam 1[b]	Exam 2[b]		Exam 1[b]	Exam 2[b]	
Dewormed areas						
School 1 ($n = 102$)	86	27	69[c]	100	87	13[c]
School 2 ($n = 94$)	42	0	100[c]	66	42	37[c]
School 3 ($n = 91$)	46	19	60[b]	52	44	15
School 4 ($n = 104$)	27	6	79[c]	59	16	72[c]
Subtotal ($n = 391$)	50	13	75[c]	70	47	32[c]
Control areas						
School 5 ($n = 100$)	87	82	6	97	94	3
School 6 ($n = 80$)	36	25	31	60	63	0
School 7 ($n = 98$)	47	38	20	51	58	0
School 8 ($n = 99$)	57	46	20	68	59	13
Subtotal ($n = 377$)	58	49	16[d]	70	69	1

[a] Schools 1, 2, 5, and 6 from urban areas, and schools 3, 4, 7, and 8 from rural areas.
[b] Exam 1 = at baseline of the study, exam 2 = 3 weeks after albendazole administration.
[c] $P = 0.0001$.
[d] $P = 0.001$.

were found in School 3 and School 7, respectively. Overall, the cure rate with albendazole was 75% (range 60–100%) for *A. lumbricoides* and 32% (range 13–72%) for *T. trichiura*. All cure rates were statistically significant for both infections except *T. trichiura* in School 3.

Table 3 shows the cure rate achieved with pyrantel pamoate administered 6 months after the start of the study. At baseline examination (exam 3), prevalence among schools varied widely, and cure rates in both *A. lumbricoides* and *T. trichiura* infections were relatively low. The cure rate in *A. lumbricoides* infections was significant only in School 1, which had the highest prevalence at baseline. As expected, the cure rate for pyrantel in *T. trichiura* infection was negligible.

Effectiveness of deworming

Figure 2 shows the prevalences of *A. lumbricoides* and *T. trichiura* infections throughout the five examinations. At baseline (exam 1), differences in prevalence of both *A. lumbricoides* and *T. trichiura* infection were significant among the four groups. Children in rural areas and covered by the programme (dewormed rural group) had the lowest prevalences of *A. lumbricoides* and *T. trichiura*. By contrast, children in urban areas who had received at least one treatment within the programme had the highest prevalences of both infections.

At 6 months, the prevalence of ascariasis had declined significantly in urban ($\chi^2 = 16$, df = 1, $P = 0.0001$) and rural ($\chi^2 = 3$, df = 1, $P = 0.0001$) dewormed areas.

Table 3. Cure rate in *Ascaris lumbricoides* and *Trichuris trichiura* infections 3 weeks after administration of pyrantel pamoate

Primary school[a]	*A. lumbricoides*			*T. trichiura*		
	% positive		Cure rate (%)	% positive		Cure rate (%)
	Exam 3[b]	Exam 4[b]		Exam 3[b]	Exam 4[b]	
Dewormed areas						
School 1 ($n = 71$)	75	61	19[c]	94	97	0
School 2 ($n = 93$)	14	8	43	34	48	0
School 3 ($n = 95$)	31	26	16	52	54	0
School 4 ($n = 97$)	17	16	6	39	44	0
Subtotal ($n = 356$)	31	25	19	52	58	0
Control areas						
School 5 ($n = 75$)	84	64	24[c]	92	85	8
School 6 ($n = 91$)	30	20	33	59	57	3
School 7 ($n = 86$)	36	44	0	57	52	9
School 8 ($n = 94$)	55	55	0	75	76	0
Subtotal ($n = 346$)	50	45	10[c]	70	67	4

[a] Schools 1, 2, 5, and 6 from urban areas, and schools 3, 4, 7, and 8 from rural areas.
[b] Exam 3 = at 6 months, before pyrantel administration, exam 4 = 3 weeks after pyrantel administration.
[c] $P < 0.05$.

Figure 2. Changes of prevalence of *Ascaris lumbricoides* and *Trichuris trichiura* at exam 1 to exam 5

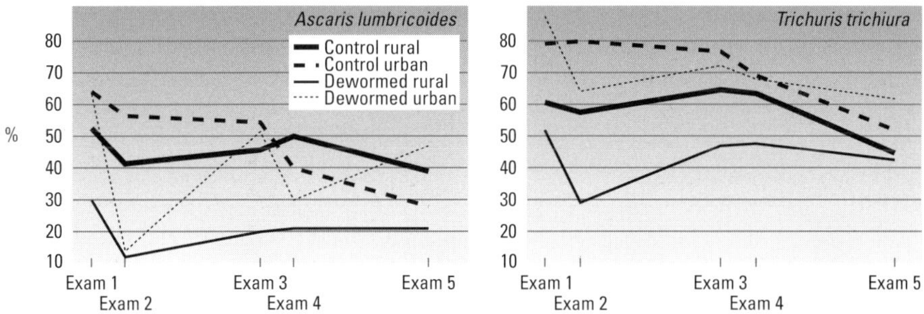

At baseline there were significant differences between urban and rural areas in the dewormed group for *A. lumbricoides* ($\chi^2 = 109$, df = 1, $P = 0.0001$) and *T. trichiura* ($\chi^2 = 130$, df = 1, $P = 0.0001$). Significant differences between urban and rural areas were also noted in the control group for *A. lumbricoides* ($\chi^2 = 5.3$, df = 1, $P = 0.021$) and *T. trichiura* ($\chi^2 = 18$, df = 1, $P = 0.0001$).

There was no significant change in the prevalence of ascariasis in either urban ($\chi^2 = 3.6$, df = 1, $P = 0.06$) or rural ($\chi^2 = 2.3$, df = 1, $P = 0.13$) control areas, and no significant reduction in trichuriasis prevalence between baseline and 6 months except in dewormed urban areas ($\chi^2 = 32$, df = 1, $P = 0.0001$).

Compared with the prevalence at baseline, however, there were highly significant reductions in the prevalence of both ascariasis and trichuriasis at 12 months ($P < 0.0001$) in dewormed groups. Unexpectedly, prevalence of both infections also fell significantly in control groups ($P < 0.001$). At the end of the year, the overall prevalences of *A. lumbricoides* and *T. trichiura* infection were 33.6% and 51.4%, respectively. The highest values were still seen in dewormed urban groups (47% and 63%, respectively, for *A. lumbricoides* and *T. trichiura*) and the lowest in dewormed rural groups (21% and 44%, respectively).

Table 4 shows changes in the intensity of ascariasis and trichuriasis over the course of the study and at exam 2 and exam 4, both of which were performed for those who had returned stool samples at the previous examination. Analyses of change in intensity between exam 2 and exam 1 and between exam 4 and exam 3 were done using paired *t*-tests.

As expected, the percentage egg reductions in *A. lumbricoides* and *T. trichiura* infections between exam 1 and exam 2 (efficacy of albendazole) were statistically significant in all dewormed groups; there was also a significant reduction in control groups for *A. lumbricoides* infection but not for *T. trichiura* infection. However, between exam 4 and exam 3 (efficacy of pyrantel), egg reduction was observed only in the dewormed urban group.

Reductions in egg count between baseline and final exams were profound, even in the control group. For *A. lumbricoides* infection, the greatest reduction was observed in the urban control group (97%) and the lowest in the rural dewormed group (44%). Similarly, for *T. trichiura* infection, the greatest reduction occurred in the urban control group (88%) and the lowest was in the rural dewormed group (55%).

Discussion

The results presented here reveal considerable heterogeneity in terms of regional and rural–urban variations in the prevalence and intensity of *A. lumbricoides* and *T. trichiura* infections in schoolchildren. They also demonstrate that deworming in Indonesia substantially reduces both prevalence and intensity of these infections in schoolchildren. Reduction of the worm burden to low levels was the most common benefit derived by children participating in the study.

It is noteworthy that prevalence and intensity of ascariasis at baseline in both South Sulawesi and West Sumatra were lower in dewormed schools than in control schools. Since the deworming programme has been running for 2 years, the initial aim of reducing prevalence of helminth infections and morbidity has been achieved. However, differences in prevalence and intensity were seen in rural areas only. Moreover, the highest prevalence and intensity were seen in urban schools in slum areas. It is to be expected that slum areas have worse sanitation and therefore higher transmission rates of infection than rural areas. In other words, reinfection rates are higher in slum areas compared with rural and other urban areas—a characteristic that should be taken into account in future programmes.

Table 4. Changes in intensity of *Ascaris lumbricoides* and *Trichuris trichiura* infections

Treatment area	Geometric mean epg[a]					Percentage egg reduction			
	Exam 1	Exam 2	Exam 3	Exam 4	Exam 5	Exam 2–1	Exam 3–1	Exam 4–3	Exam 5–1
				Ascaris lumbricoides					
Dewormed									
Rural	9 (441)	1 (194)	4 (450)	4 (192)	5 (424)	92[b]	56[b]	0	44[b]
Urban	224 (445)	3 (196)	89 (413)	14 (164)	63 (352)	99[b]	60[b]	84[b]	72[b]
Subtotal	45 (886)	2 (391)	18 (863)	7 (356)	15 (776)	96[b]	60[b]	61[b]	67[b]
Control									
Rural	60 (220)	25 (197)	43 (221)	59 (180)	19 (212)	52[b]	28	–37	68[b]
Urban	265 (217)	156 (180)	132 (202)	37 (166)	9 (182)	50[b]	50	72	97[b]
Subtotal	125 (437)	61 (377)	73 (423)	47 (346)	13 (394)	50[b]	42	36	90[b]
				Trichuris trichiura					
Dewormed									
Rural	31 (441)	5 (195)	16 (450)	16 (192)	14 (424)	35[b]	48[b]	0	55[b]
Urban	652 (445)	123 (196)	228 (413)	155 (164)	72 (352)	76[b]	65[b]	32	89[b]
Subtotal	141 (886)	27 (391)	57 (863)	46 (356)	29 (776)	80[b]	60[b]	19	79[b]
Control									
Rural	46 (220)	42 (197)	63 (221)	62 (180)	14 (212)	2.3	–37	2	70[b]
Urban	340 (217)	419 (180)	284 (202)	161 (166)	40 (182)	0.7	17	43	88
Subtotal	125 (437)	127 (377)	130 (423)	98 (346)	22 (394)	1.6	–4	25	82[b]

[a] Sample size, *n*, in parentheses.
[b] $P < 0.05$. *P* values were obtained from independent *t*-tests (exam 3–1 and exam 5–1) and paired *t*-tests (exam 2–1 and exam 4–3).

The prevalence of trichuriasis in the dewormed group in West Sumatra was significantly higher than that in control group, with the highest levels in schools in urban—especially slum—areas. This may be a function of sanitary conditions rather than a reflection of the efficacy of albendazole. The lower prevalence of trichuriasis in the dewormed group in rural areas of West Sumatra resulted from the administration of albendazole. On the other hand, it was observed that sanitary conditions for the dewormed group in slum areas of West Sumatra were worse than those of the control group in slum areas. It might therefore be expected that the transmission rate of trichuriasis was higher among the dewormed group than in the control group.

As expected, the efficacy of albendazole in treating *A. lumbricoid*es infections was relatively high at 75% (range 60–100%). Only one school had a 100% cure rate; in the other three schools, cure rates were below 80%. This finding is not in agreement with the results of several other studies in Indonesia, reviewed by Abidin (1993), which found a higher cure rate with albendazole (95–100%). In addition, results from a study in Malaysian schoolchildren (Rahman, 1996) showed a cure rate of 87%. The lowest cure rate in the present study (60%) was found in a school located in a rural area (close to the city) where the initial prevalence of *A. lumbricoides* was 46%. These differences may be a result of the sample size, specific situations (e.g. children refusing—or spitting out—the tablets), or the quality of the drug. As mentioned, the albendazole used in this study was a generic formulation produced locally. It may be necessary to undertake an efficacy study comparing the locally produced drug with a proprietary preparation.

The efficacy of albendazole against *T. trichiura* in this study proved to be low—32%, varying from 13% to 72%. This accords with results from studies in other places in Indonesia (Abidin, 1993), which have shown a cure rate of 27–59%. In Malaysian schoolchildren, Rahman (1996) reported a higher cure rate for albendazole against *T. trichiura* infection (83%). Again, differences may be due to sample size, host condition, or the quality of the drug.

In the study of pyrantel pamoate efficacy against *A. lumbricoides*, a lower cure rate was observed than in an earlier study in slum areas of Makassar (Hadju et al., 1996b). The mean cure rate in the present study was 19%, with a range from 6% to 43%. The apparent efficacy of this drug may relate to the initial prevalence of the infection: only in School 1, where initial prevalence was high (75%), was there a significant reduction in prevalence. Possible reasons for this low cure rate include incorrect drug administration and inadequate drug quality. For the most part, the drug was administered by field workers directly to pupils at school, so incorrect administration seems unlikely; only a few children received the drug at home. There is no information on the quality of the drug. Problems of drug resistance cannot be entirely ruled out, although there are no grounds yet for supposing that any resistance has occurred.

It is interesting to note that the prevalence and the intensity of both *A. lumbricoides* and *T. trichiura* infections in control areas decreased significantly, particularly in the last 6 months of the study. Similar observations in control groups have been made in other community-based studies (Hadju et al., 1996b). A study in slum areas found significant reductions in all groups at the end of the 12-month study (Hadju et al., 1997). Several explanations for this have been proposed, including seasonal variation (Cabrera, 1998), psychological effects on the untreated group of treatment of other people (Thein-Hlaing et al., 1990, 1991), variability of egg counts in faeces (Hall, 1981), and receiving anthelminthic treatment from other sources. This last factor may have applied to this study, since parents of children in target schools received information about the effects of helminth infection on their children's health. Moreover, anthelminthic drugs are available at most local health centres and in local stores and pharmacies.

Declining trends in the prevalence and intensity of both infections in each group seem similar. Groups with higher prevalences at baseline also had higher prevalences at the end of the study, and groups with lower prevalences on first examination also had lower prevalences at the final examination. National deworming programmes should not ignore environmental behavioural factors (Holland, 1989): a nationwide programme in Indonesia would need to consider the huge diversity in the characteristics of the territories and communities of the country. Underlying factors such as lack of hygiene and poor sanitation should be addressed simultaneously in a national programme.

This study concludes that deworming through the school feeding programme in Indonesia has substantially reduced helminth infection in treated areas and may have had useful effects in other areas. Further work is needed to examine the quality of drugs, to eliminate possible quality control problems as a source of the unexpectedly low cure rates of anthelminthic drugs used in the programme. Careful consideration should be given to the continuation of the programme in areas with high transmission rates (such as slum areas). In addition, the need for monitoring and evaluation of future deworming activities should be emphasized.

Acknowledgements

The study was financially supported by the Centre for School Feeding Project, Indonesia. The authors wish to thank Vivian Krause for her help in editing and for providing valuable comments on the manuscript. They also thank all the schoolchildren and parents involved in the study for their willing participation.

References

Abidin SAN (1993) Albendazole pada pengobatan Nematoda usus. [Treatment of albendazole on helminthiases.] *Majalah Parasitologi Indonesia*, 6:756–782.

Adams EJ et al. (1994) Physical activity and growth of Kenyan schoolchildren with hookworm, *Trichuris trichiura*, and *Ascaris lumbricoides* infections are improved after treatment with albendazole. *Journal of Nutrition*, 124:1199–1206.

Bundy DAP, Kan SP, Rose R (1988) Age-related prevalence, intensity and frequency distribution of gastrointestinal helminth infection in urban slum children from Kuala Lumpur, Malaysia. *Transactions of the Royal Society of Tropical Medicine and Hygiene*, 82:289–294.

Bundy DAP et al. (1990) Control of geohelminths by delivery of targeted chemotherapy through schools. *Transactions of the Royal Society of Tropical Medicine and Hygiene*, 84:115–120.

Cabrera B (1998) Seasonal variation of *Ascaris* infection in the Philippines and the determination of appropriate interval of mass treatment. In: *Proceedings of IX International Congress of Parasitology (ICOPA), August 1998, Chiba, Japan*. Bologna, Monduzzi Editore:133–138.

Hadidjaya P et al. (1996) The effect of ascariasis treatment and health education on the cognitive function of primary school children. *Medical Journal of Indonesia*, 5:189–194.

Hadju V et al. (1995) Intestinal helminthiasis, nutritional status, and their relationship: a cross-sectional study in urban slum school children in Indonesia. *Southeast Asian Journal of Tropical Medicine and Public Health*, 26:719–729.

Hadju V et al. (1996a) Improvements in appetite and growth in helminth-infected schoolboys three and seven weeks after a single dose of pyrantel pamoate. *Parasitology*, 113:497–504.

Hadju V et al. (1996b) Comparison between albendazole and pyrantel pamoate once and twice yearly in urban slum school children in Ujung pandang (Makassar). *Medical Journal of Indonesia*, 5:195–202.

Hadju V et al. (1997) Relationships between soil-transmitted helminthiases and growth in urban slum schoolchildren in Ujung Pandang, Indonesia. *International Journal of Food Science and Nutrition*, 48:85–93.

Hall A (1981) Quantitative variability of nematode egg counts in faeces: a study among rural Kenyans. *Transactions of the Royal Society of Tropical Medicine and Hygiene*, 75:682–687.

Holland CV (1989) Man and his parasites: integration of biomedical and social approaches to transmission and control. *Social Science and Medicine*, 29:403–411.

Jalal F et al. (1998) Serum retinal concentrations in children are affected by food sources of beta-carotene, fat intake, and anthelminthic drug treatment. *American Journal of Clinical Nutrition*, 68:623–629.

Karyadi E et al. (1996) Anthelminthic treatment raises plasma iron levels but does not decrease the acute-phase response in Jakarta school children. *Southeast Asian Journal of Tropical Medicine and Public Health*, 27:742–753.

Latham MC et al. (1990) Metrifonate or praziquantel treatment improves physical fitness and appetite of Kenyan schoolboys with *Schistosoma haematobium* and hookworm infection. *American Journal of Tropical Medicine and Hygiene*, 43:170–179.

Nokes C et al. (1991) Geohelminth infection and academic assessment in Jamaican children. *Transactions of the Royal Society of Tropical Medicine and Hygiene*, 85:272–273.

Nokes C et al. (1992) Moderate to heavy infections of *Trichuris trichiura* affect cognitive function in Jamaican school children. *Parasitology*, 104:539–547.

Pegelow K et al. (1997) Parasitological and nutritional situation of school children in the Sukaraja District, West Java, Indonesia. *Southeast Asian Journal of Tropical Medicine and Public Health*, 28:173–190.

Rahman WA (1996) Comparative trials using albendazole and mebendazole in the treatment of soil-transmitted helminthiases in schoolchildren on Penang, Malaysia. *Southeast Asian Journal of Tropical Medicine and Public Health*, 27:765–767.

Simeon T et al. (1994) School performance, nutritional status, and *trichuriasis*. *Acta Paediatrica*, 83:1188–1193.

SPSS (1995) *SPSS base system syntax reference guide, Release 7.5*. Chicago, SPPS Inc.

Stephenson LS et al. (1993) Physical fitness, growth, and appetite of Kenyan schoolboys with hookworm, *Trichuris trichiura*, *Ascaris lumbricoides* infections are improved four months after a single dose of albendazole. *Journal of Nutrition*, 123:1036–1046.

Thein-Hlaing, Than-Saw, Myat-Lay-Kyin (1990) Control of ascariasis through age-targeted chemotherapy: impact of 6-monthly chemotherapeutic regimes. *Bulletin of the World Health Organization*, 68:747–753.

Thein-Hlaing et al. (1991) The impact of three-monthly age-targeted chemotherapy on *Ascaris lumbricoides* infection. *Transactions of the Royal Society of Tropical Medicine and Hygiene*, 85:519–522.

WHO (1991) *Basic laboratory methods in medical parasitology.* Geneva, World Health Organization.

WHO (1992) Health of school children: treatment of intestinal helminths and schistosomiasis. Geneva, World Health Organization (document WHO/CDS/IPI/CTD/92.1).

WHO (1992) *Prevention and control of parasitic infections. Report of a WHO Expert Committee.* Geneva, World Health Organization (WHO Technical Report Series, No. 749).

Practical experience from the primary school-based soil-transmitted helminth control programme in Jakarta (1987–1999)

Adi Sasongko,[1] Firman Lubis, Mundi Mahaswiati, Dewi Ratnawulan
Yayasan Kusuma Buana, Jakarta, Indonesia

Introduction

Intestinal worm infection has been recognized as a major health problem in Indonesia since the Dutch colonial period: the history of public health in the country records that a hookworm control project was initiated by the Dutch colonial government in 1930 in Banyumas, Central Java, in collaboration with the Rockefeller Foundation. Later, in 1975, a nationwide project to control intestinal worm infections was started by the Government of Indonesia but discontinued in the 1980s because of other health priorities and budgetary limitations (Adhyatma, 1997[2]). Since then, a number of sporadic and small-scale projects have been implemented but then discontinued after a short period. As a result, the prevalence of intestinal worm infections remains very high (P4I, 1992; Ministry of Health, 1995; Sasongko, Mahaswiati & Lubis, 1997;[3] Margono & Ismid, 1998;[4] Margono et al., 1999; Sasongko, Ratnawulan & Lubis, 1999; Sasongko, 1999[5]). The long-term adverse impact this will have on the health and nutritional status of Indonesia's population should not be underestimated.

Learning from Japan's successful experiences in controlling worm infection through primary schools (Kunii, 1992), Yayasan Kusuma Buana (YKB) initiated a school parasite control programme in 1987 in collaboration with the Japan

---

1 Dr A. Sasongko, Jalan Asem Baris Raya Blok A-3, Tebet, Jakarta Selatan 12430, Indonesia
Tel: +62 21 861 6723; Fax: +62 21 831 4764; E-mail: ykb-jkt@idola.net

2 Adhyatma (1997) *Berbagai model pemberantasan cacingan di Indonesia. [Different models of STH control in Indonesia.]* Presented at the Seminar on One Decade of STH Control in Jakarta, organized by Faculty of Public Health University of Indonesia and Yayasan Kusuma Buana, 21 October, 1997.

3 Sasongko A, Mahaswiati M, Lubis F (1997) Hasil pemberantasan cacingan di sekolahssekolah dasar DKI Jakarta 1987–1997. [The results of STH control in primary schools of Jakarta 1987–1997.] Presented at the Seminar on One Decade of STH Control in Jakarta, organized by Faculty of Public Health, University of Indonesia, and Yayasan Kusuma Buana, 21 October 1997.

4 Margono Sri S, Ismid IS (1998) *Mass treatment of soil-transmitted helminthoses in schoolchildren in Indonesia.* Presented at IX International Congress of Parasitology, 24–28 August, 1998, Chiba, Japan.

5 Sasongko, A (1999) *Duabelas tahun pelaksanaan program pemberantasan cacingan di sekolah-sekolah dasar DKI Jakarta, 1987–1999. [Twelve years' implementation of primary school-based parasite control in Jakarta, 1987–1999.]* Presented at the VIIth National Congress of Indonesian Parasite Control Association, Makassar, 18–21 November 1999.

Association for Parasite Control (JAPC) and the Japan Organization for International Cooperation in Family Planning (JOICFP). During the period 1987–1991, experts from JAPC and JOICFP provided technical and financial assistance with the establishment of this innovative programme; after 1991, YKB continued its activities independently and expanded the programme from 34 schools in 1987 to more than 500 schools in 1999. The Jakarta Chapter of PKBI (Indonesian Planned Parenthood Association) was also part of the programme from 1987 to 1995, but then discontinued their involvement because of changed priorities.

Objective

The objective of the YKB School Parasite Control Programme is to control intestinal worm infection among primary-school children in Jakarta and to promote a healthy life style (in terms of cleanliness and personal hygiene) in the school community.

Approach

The programme relies on a school-based preventive approach to infection control, involving:

— health education activities (aimed at pupils, teachers, and parents)
— regular (twice yearly) stool examination
— screening-based treatment (twice yearly)
— other health promotion activities such as school cleanliness competitions, and drawing and writing competitions.

Since its inception in 1987, the programme has adopted a self-reliant approach to sustaining its activities. A contribution of Rp. 1.000[1] per pupil is collected from parents each year towards the cost of two stool examinations and any treatment required.

The programme is carried out by YKB's network of five maternal and child health and family planning clinics in Jakarta, and all stool examinations are undertaken in the central laboratory of YKB. A tripartite collaboration of YKB with related government agencies (Jakarta health and education authorities) and the medical school at the University of Indonesia (Department of Parasitology) provides support for the programme. Further support comes from collaboration with private industry. Since 1991, Pfizer Indonesia has provided anthelminthics (Combantrin and later on Quantrel) at a discounted price, plus information, education, and communication (IEC) materials for distribution to the school community and sponsorship for health promotion activities.

[1] The exchange rate for US$1 was Rp. 1.800 in 1987 and about Rp. 7.5000 in 2000.

Geographically organized approach to achieving mass coverage

Because of limited resources, the programme has expanded gradually. Initially, only schools in the immediate vicinity of YKB clinics were covered by the programme; 4–5 schools in each location (kelurahan or village) were involved in the first year, and another 5–10 new schools were added in subsequent years. After all local schools were covered, the programme expanded to cover schools in the neighbouring kelurahans.

Thus, a point was gradually reached at which almost all primary-school children in each kelurahan were covered by the programme. Twice-yearly stool examinations were carried out, followed by treatment of positive cases with anthelminthics. With infection control on this scale, there is a significant decrease in worm egg contamination (particularly infective fertile eggs) of the surrounding area by the previously infected children. Clearly, this reduces the potential for reinfection.

A cohort analysis of a group of 40 schools, carried out by YKB over the period 1988–1991, showed that after 3 consecutive years of control activities the ratio of fertile to infertile *Ascaris* eggs would be inverted. Even though soil contamination might continue for some time, infertile eggs—the predominant contaminants—are not infective. The impact of the reduced number of fertilized eggs contaminating the soil enhances the effectiveness of a geographically organized approach to mass coverage.

Establishment of a mass-capacity stool examination laboratory to support a screening-based approach

To support the screening-based approach used in this programme, there should be regular examination of stool samples from all pupils. Starting in 1987, YKB gradually established a mass-capacity screening laboratory. Six technicians were given a basic training by the Department of Parasitology, Faculty of Medicine, University of Indonesia; further on-the-job training was given by JAPC during 1987–1990.

Stool samples are examined by the cellophane thick-smear (Kato–Katz) method (WHO, 1991) to identify and differentiate the eggs of *Ascaris lumbricoides*, *Trichuris trichiura*, and hookworm (in Indonesia, *Necator americanus*). Semi-quantitative measurement uses a + to +++ scale based on the number of eggs found (Vajrasthira & Waikagul, 1999).[1] *Ascaris* eggs are further categorized as fertile or infertile.

The process of stool examination is organized into several steps:

1. A stool container is issued to every pupil.
2. Stool specimens are collected from the school.

[1] + indicates 1–9 eggs, ++ indicates 10–99 eggs, +++ indicates 100 or more eggs.

3. Specimens are sorted according to school, grade, and alphabetical order of pupils' names.
4. Specimens are stored in a refrigerator if they cannot be examined on the day of collection.
5. Stool specimens are prepared for microscopic examination (two specimens on one slide).
6. Specimens are dried (for 10–15 minutes).
7. Specimens are examined under a binocular microscope, and result are recorded.
8. Results are entered into the computer.
9. Results are printed out and sent to the clinics and schools.

The microscopic examination (step 7) is done exclusively by the laboratory technicians; the rest of the process is completed by other laboratory staff. Using an "assembly line" approach, one laboratory technician can examine up to 300 specimens a day. At present, YKB has four senior technicians, two junior technicians, and a number of other laboratory staff.

After results are reported back to the clinic, clinic and laboratory staff visit the school to oversee the administration of anthelminthics by the teachers.

Treatment

During the first 4 years of the programme, all positive cases were treated with pyrantel pamoate, 10 mg/kg body weight. However, the laboratory results showed a significant effect only in *Ascaris* infection. In the fifth year, therefore, treatment was changed to a combination of pyrantel and oxantel pamoate, which resulted in a significant reduction in *Ascaris, Trichuris,* and hookworm.

As stated, treatment is carried out in schools. Teachers gather the children identified as infected and administer the anthelminthics directly under the supervision of the clinic/laboratory staff. A printed form letter giving the results of the earlier stool examination is distributed to all parents before the day of treatment. Stool examination itself is scheduled at least 3 months after the previous treatment. As the life cycle of the worms is about 3 months, this period makes it possible to determine whether reinfection has occurred since the previous treatment. The ability to track the pattern of infection is one of the benefits of the screening approach.

Results

During the 12 years since 1987, the programme's coverage increased from 34 schools with 9591 students to 507 schools with 124981 schoolchildren. Total prevalence rate of intestinal worm infection in the covered schools declined from 78.6% to 11.7%; ascariasis declined from 62.2% to 5.9%, trichuriasis from 64.7% to 6.5%, and hookworm from 1.4% to 0.02% (Figures 1–4). A total of 1 428 695 stool specimens have been examined and 555 695 doses of anthelminthics have been given selectively to infected children (see Table 1).

Table 1. Results of primary school-based soil-transmitted helminth control programme in Jakarta, 1987–1999

Year	Cycle	No. of schools	No. of students	No. of lab. exams	%	Pos. cases	%	Asc.	%	Trich.	%	Hkw.	%	Oxy.	%
1987/88	A	34	9591	7879	82.1	6193	78.6	4902	62.2	3793	48.1	57	0.72	0	0.00
	B	34	9786	6586	67.3	4670	70.9	1117	17.0	4260	64.7	44	0.67	0	0.00
1988/89	C	87	25176	17380	69.0	11906	68.5	7453	42.9	9707	55.9	244	1.40	63	0.36
	D	97	27617	17441	63.2	11195	64.2	5276	30.3	9463	54.3	203	1.16	104	0.60
1989/90	E	208	56034	37974	67.8	25722	67.7	15919	41.9	20926	55.1	417	1.10	186	0.49
	F	196	54596	27101	49.6	18567	68.5	7543	27.8	16797	62.0	325	1.20	189	0.70
1990/91	G	388	103750	67716	65.3	47224	69.7	27877	41.2	40069	59.2	432	0.64	334	0.49
	H	400	109065	53618	49.2	32985	61.5	12469	23.3	28675	53.5	381	0.71	284	0.53
1991/92	I	553	154026	91401	59.3	61221	67.0	34892	38.2	52117	57.0	560	0.61	125	0.14
	J	542	152905	67269	44.0	37740	56.1	15141	22.5	32741	48.7	378	0.56	227	0.34
1992/93	K	651	176680	100516	56.9	56562	56.3	28937	28.8	45863	45.6	349	0.35	271	0.27
	L	616	168033	70082	41.7	28555	40.7	13745	19.6	21601	30.8	234	0.33	237	0.34
1993/94	M	586	150436	85524	56.9	36604	42.8	19908	23.3	27838	32.5	243	0.28	342	0.40
	N	538	148899	73463	49.3	22922	31.2	12688	17.3	15742	21.4	148	0.20	159	0.22
1994/95	O	569	140871	84368	59.9	30263	35.9	17299	20.5	21672	25.7	97	0.11	250	0.30
	P	510	128141	66332	51.8	17798	26.8	8145	12.3	12729	19.2	105	0.16	142	0.21
1995/96	Q	535	130401	74879	57.4	19307	25.8	9943	13.3	13000	17.4	50	0.07	190	0.25
	R	527	131380	67475	51.4	14256	21.1	6526	9.7	9492	14.1	80	0.13	158	0.23
1996/97	S	518	122518	74522	60.8	18193	24.4	8595	11.5	12459	16.7	38	0.05	162	0.22
	T	508	121504	60904	50.1	11069	18.2	4937	8.1	7218	11.9	50	0.08	130	0.21
1997/98	U	562	137933	82584	59.9	16649	20.2	10082	12.2	9968	12.1	14	0.02	76	0.09
	V	495	126025	64209	50.9	8829	13.8	4080	6.4	5653	8.8	5	0.01	37	0.06
1998/99	W	515	124432	72005	57.9	10559	14.7	5410	7.5	6350	8.8	16	0.02	63	0.09
	X	507	124981	57467	46.0	6706	11.7	3406	5.9	3724	6.5	12	0.02	45	0.08

Total number of laboratory examinations: 1428695; total number of students treated: 555695.

Abbreviations:

Cycle	cycle of examination (twice a year)
No. of lab. exams	number of laboratory examinations
Pos. cases	positive cases of worm infection
Asc.	*Ascariasis lumbricoides* (roundworm)
Trich.	*Trichuris trichiura* (whipworm)
Hkw.	hookworms
Oxy.	*Oxyuris vermicularis* (pinworm)

Figure 1. Declining prevalence of worm infection in primary schools of Jakarta, 1987–1999

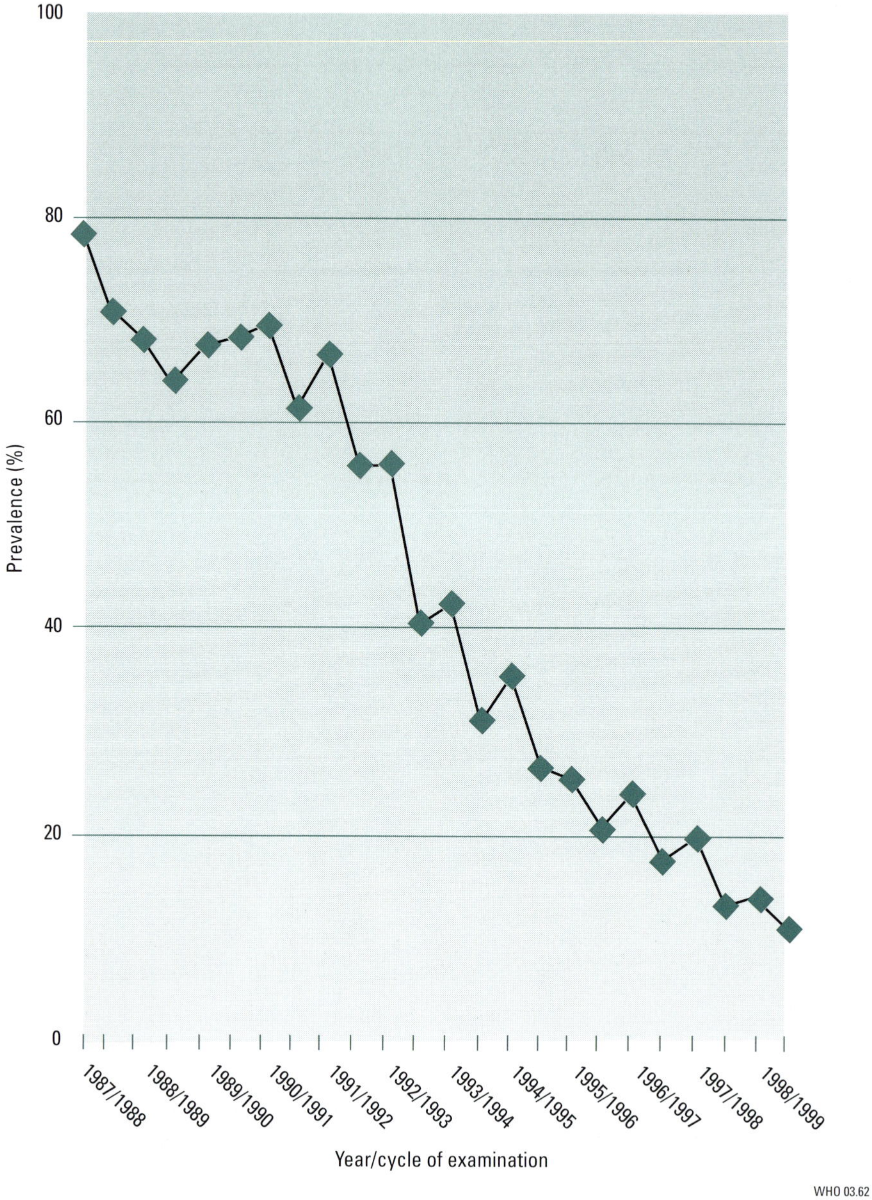

To encourage a healthy lifestyle among schoolchildren and in the school community generally, educational activities have targeted not only the children but also the teachers and parents. Health education classes have been organized to explain: the impact of intestinal worm infection on health, nutrition, and learning capacity; how worm infection can be prevented; and—for the parents and

Figure 2. Declining prevalence of ascariasis in primary schools of Jakarta, 1987–1999

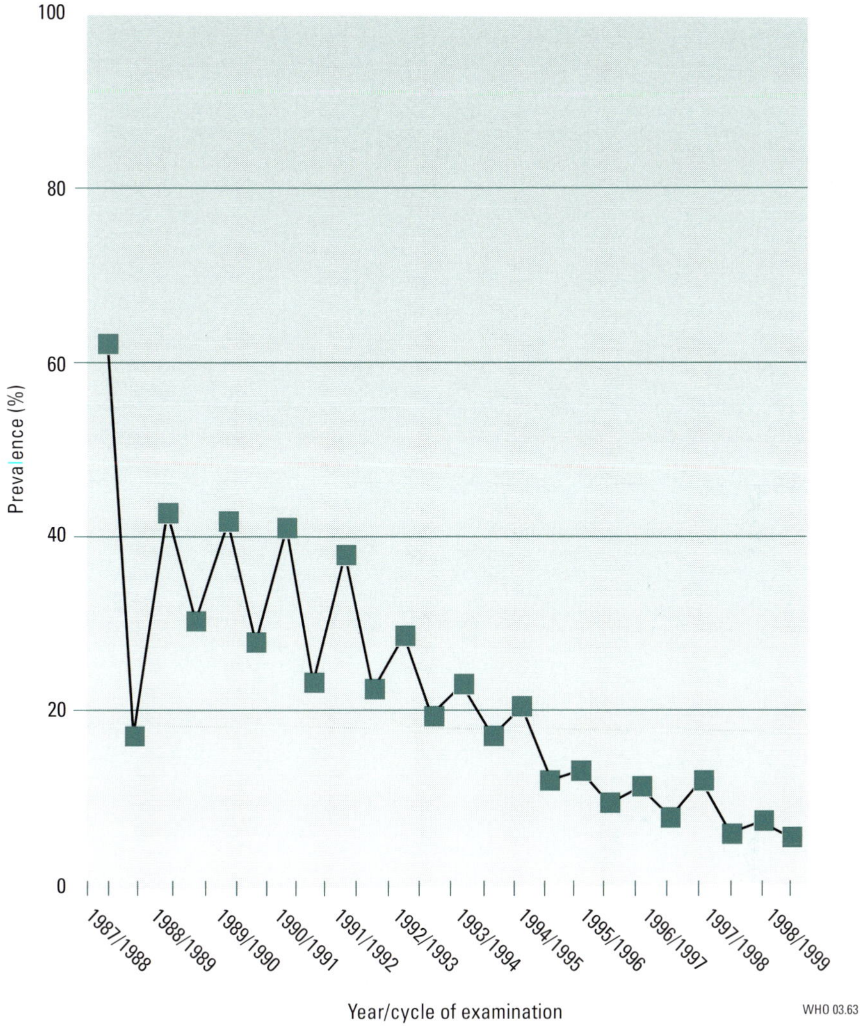

teachers—the purpose and nature of the programme's activities. Printed educational materials, such as leaflets and posters, have been distributed.

"Snowball" growth of the programme

After implementation of the programme in the schools close to YKB's clinics, more and more personal contacts developed between the clinic's staff, the teachers, and the parents. As a result, the surrounding community became increasingly aware of the clinics' services—and clinic data show an increase in visits after the establishment of the helminth programme in 1987. Frequent personal contact with the teachers also enables YKB to identify teachers' needs: in response to

Figure 3. Declining prevalence of trichuriasis in primary schools of Jakarta, 1987–1999

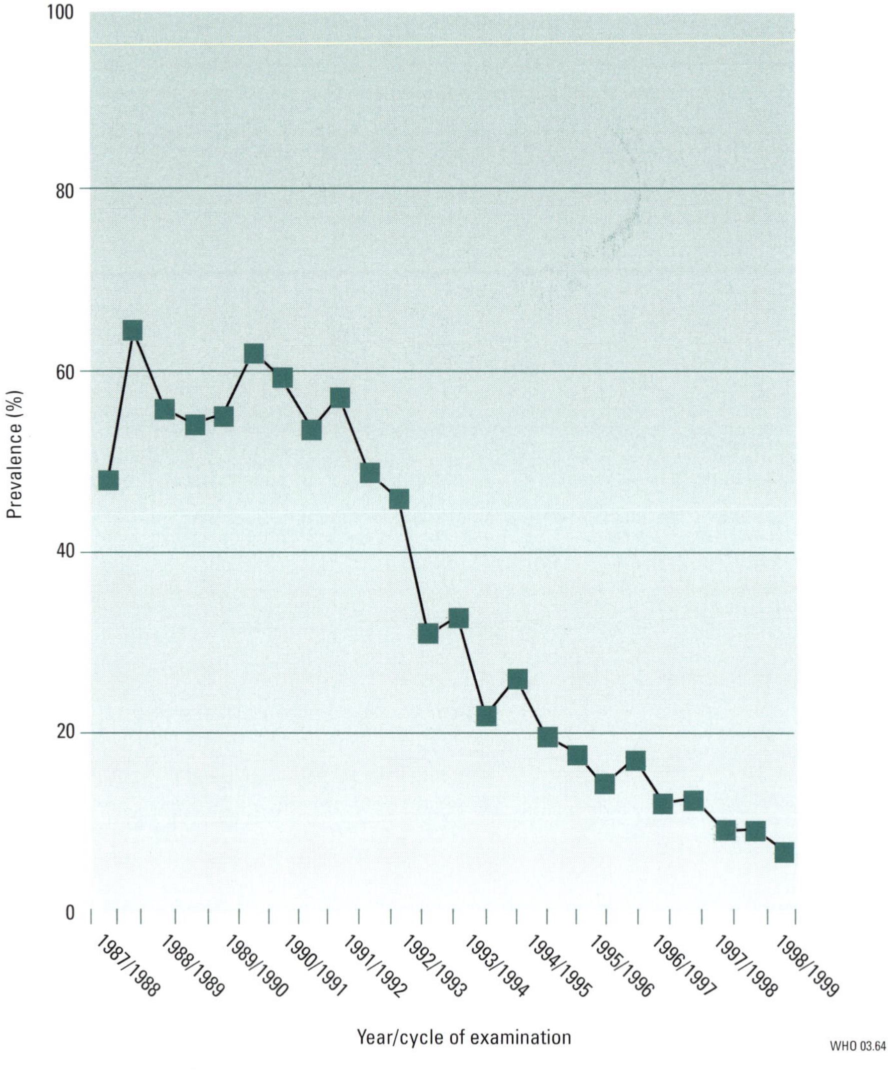

teachers' requests, YKB's clinics have provided cervical smear examinations since 1991. Between 1993 and 1999, a series of 17 one-day "Health and Education" seminars were held to discuss teaching methods and health and nutrition topics. Each was attended by an average of 250 teachers (Sasongko, 1999[1]).

[1] Sasongko A (1999) *Duabelas tahun pelaksanaan program pemberantasan cacingan di sekolah-sekolah dasar DKI Jakarta, 1987–1999. [Twelve years' implementation of primary school-based parasite control in Jakarta, 1987–1999.]* Presented at the VIIth National Congress of Indonesian Parasite Control Association, Makassar, 18–21 November 1999.

Figure 4. Declining prevalence of hookworm infection in primary schools of Jakarta, 1987–1999

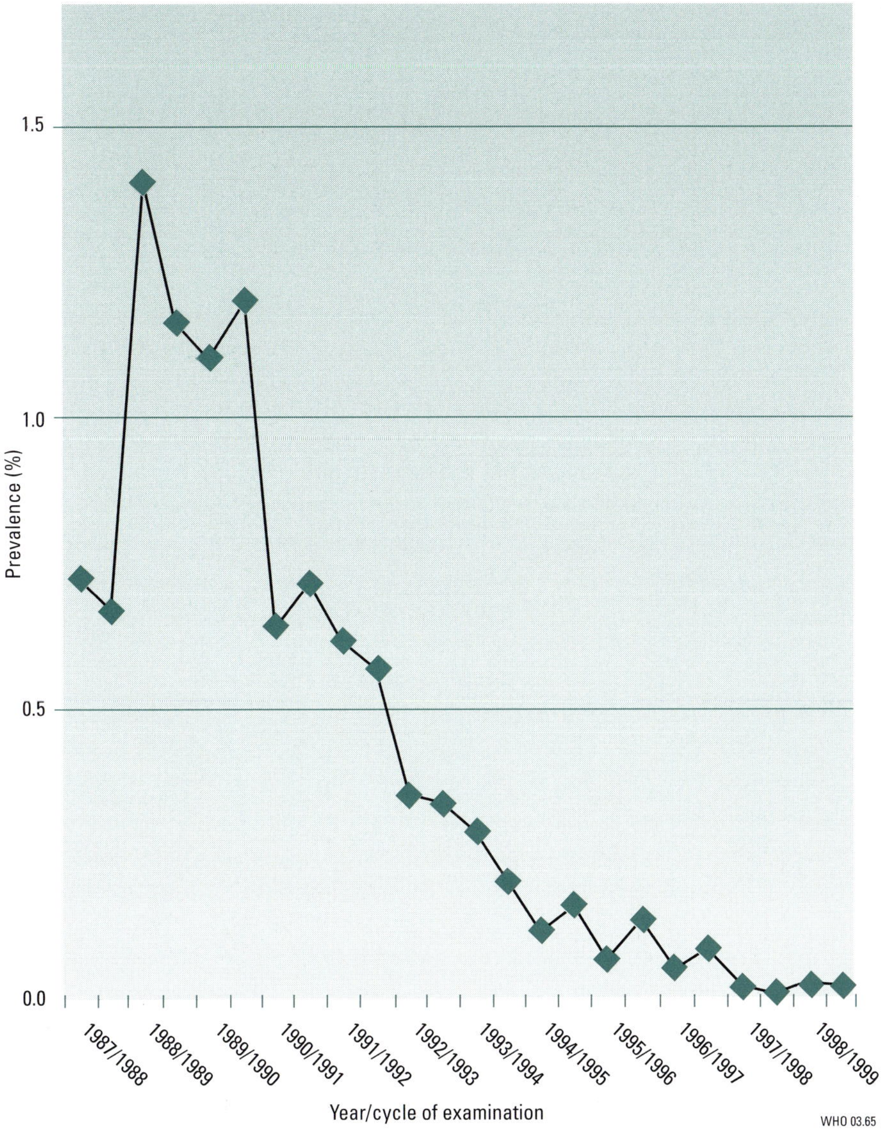

All the above-mentioned developments have been implemented on a "self-reliance" basis, with teachers contributing towards the costs. Attendance at a one-day seminar in 1999 cost Rp. 15.000; the fee for a cervical smear examination was Rp. 25.000.

The "snowball" growth of activities is clear. What was once a school parasite control programme has now become a more comprehensive school health pro-

Table 2. Distribution of STH eggs counts according to intensity in Panggang and Pramuka Islands, North Jakarta, 4 August 1999

Infection intensity[a]	*Ascaris lumbricoides*		*Trichuris trichiura*		Hookworm	
	No.	%	No.	%	No.	%
+	157	29.2	690	99.0	0	0.0
++	18	3.4	7	1.0	0	0.0
+++	0	0.0	0	0.0	0	0.0
(+)	188	35.0	0	0.0	0	0.0
(++)	161	30.0	0	0.0	0	0.0
(+++)	13	2.4	0	0.0	0	0.0
Total	537		697		0	

[a] + = 1–9 eggs, ++ = 10–99 eggs, +++ = 100 or more eggs; (+), (++), (+++) = numbers of infertile eggs.

gramme, targeting teachers as well as the schoolchildren. YKB's reputation for expertise in intestinal worm control has also grown, with the result that stool examination requests have been received from other organizations such as the Population Council of Indonesia, Nordic Women's Association, a number of commercial establishments, the health authority of Tangerang, West Java, and the Coordinating Body for Social Welfare in Jakarta (Ratnawulan, 2000).

From 1997, YKB expanded its activities to the Seribu Islands north of Jakarta. During the first 2 years, activities were directed towards the community because YKB's team was able to visit the island only during the school vacation in June. In August 1999, the team visited the islands during school days and the focus of activities switched to the school community. At this time, laboratory equipment was taken to the islands so that stool examinations could be conducted on site and treatment provided as soon as results were available. Important data were gathered not only from primary-school children but also from pupils in junior and senior high schools. Surprisingly high prevalence rates (see Tables 2 and 3) were found among both primary-school children (95.1%) and high-school pupils (86.7%); these were reported to the Vice-Governor of Jakarta and the Jakarta health and education authorities (Sasongko, Mahaswiati & Lubis, 1997;[1] Sasongko, Ratnawulan & Lubis, 1999).

Community-self-financing

The best method for controlling soil-transmitted helminth (STH) infection is to establish a long-term programme—and such a programme should not have to be dependent on external funds. Sustainable activities become possible with the par-

[1] Sasongko A, Mahaswiati M, Lubis F (1997) *Hasil pemberantasan cacingan di sekolahssekolah dasar DKI Jakarta 1987–1997. [The results of STH control in primary schools of Jakarta 1987–1997.]* Presented at the Seminar on One Decade of STH Control in Jakarta, organized by Faculty of Public Health, University of Indonesia, and Yayasan Kusuma Buana, 21 October 1997.

Table 3. Prevalence rate of intestinal worm infections based on target groups on Panggang and Pramuka Islands, North Jakarta, 4–8 August 1999

Items	Target groups							
	Total		Primary school		High school		General community	
	n	%	*n*	%	*n*	%	*n*	%
Total container	907		565		269		73	
Empty container	18		13		5		0	
Total specimen	889	100.0	552	100.0	264	100.0	73	100.0
Positive cases	811	91.2	525	95.1	229	86.7	57	78.1
Ascaris lumbricoides	537	60.4	361	65.4	142	53.8	34	46.6
Trichuris trichiura	697	78.4	476	86.2	177	67.0	44	60.3

ticipation of the community. To sustain its parasite control activities (within and outside schools), YKB uses a fee-charging approach; the money collected is channelled back to the community to finance affordable, good-quality services.

During the period 1987–1999, parents contributed Rp. 679 176.680 to partly finance school parasite control activities. Teachers' contributions to financing cervical smear examinations and continuing educational activities totalled Rp. 92 500.000.

Constraints to implementation

A number of constraints were revealed during implementation of this programme. Regular health checks are not common in Indonesia: people believe that they need a health examination only once they are sick. The cost of health examinations reinforces this belief. YKB encountered similar difficulties in promoting regular stool examination of schoolchildren, even when they provided the services directly at schools and charged little. Not all pupils regularly and reliably submitted stool specimens to the YKB field workers. There were similar problems in collecting parents' contributions. Some parents still prefer using their money for purposes other than paying the contribution.

Limited resources have meant that expansion of the programme's coverage has been slow. In the beginning, participating schools were located in the immediate vicinity of YKB's clinics and laboratory. Since then, the greater the coverage, the further the distance from YKB's facilities to the schools and the higher the cost of providing the services. The fact that the contribution fee has remained unchanged since 1987 has also limited the rate at which coverage has expanded.

Other STH control projects implemented in Jakarta add to the difficulty of managing this programme. Where neighbouring schools offer projects that are free of charge and distribute anthelminthics without stool examinations it becomes more difficult to foster acceptance of fee-charging, screening-based treatment.

Practical experiences: lessons from the field

1. Controlling helminth infection is a long-term "battle". Long-term intervention requires gradual, rather than abrupt, expansion of programme activities. At the same time, the ability to manage a mass-coverage programme needs to be gradually learned and mastered.
2. Sustainable intervention needs a strong foundation—and the best foundation is community support. Initial capital from a donor agency can be used to start a programme but the principle of community support should be established from the outset.
3. The charging of reasonable fees should be developed as a form of community support in order to sustain the activities.
4. Establishing an STH control programme as a separate and exclusive entity is expensive. Integrating the programme with existing infrastructure in a synergistic manner is a much more cost-effective option.
5. The control programme should aim to achieve mass coverage gradually, expanding into areas surrounding those that are already covered. This approach reduces the number of infective eggs contaminating the soil, and results in a profound drop in the rate of reinfection as well as in total prevalence.
6. Results of stool examinations should be communicated to teachers and parents, whether they are positive or negative or simply reflect the fact that children were not submitting stool sample for examination. This regular feedback mechanism helps to maintain support for the programme by keeping the community informed about the activities. Awareness of the children's health status also provides the motivation for adopting a healthy lifestyle and so helps to prevent worm infection.
7. In the long-term, a screening-based approach is more cost-effective than blanket coverage. The economies of scale will lower the cost of laboratory examinations and the declining prevalence will reduce the expenditure on anthelminthics. Regular reporting of stool examination results helps to motivate pupils (as well as parents and teachers) to improve their cleanliness and personal hygiene in order to prevent reinfection. Limiting treatment to positive cases also teaches all parties that drugs should not be taken without clear indication of infection.
8. The urban community is more likely than others to accept the charging of fees in a screening-based STH control programme. Once the programme is established in urban areas, it can then be expanded to surrounding suburban and rural areas.

9. STH control could be developed as an entry point for the provision of more comprehensive school health services, including services for the teachers.
10. A consistent decline in STH prevalence was observed despite the overall poor sanitation conditions in Jakarta—particularly in the areas surrounding participating schools and residential areas. Thus, even without the substantial expenditure required for physical improvements in sanitation, it is still possible to control intestinal worm infection. Nonetheless, improved sanitation conditions would certainly accelerate the decline in prevalence.
11. The control of STH can be achieved only through a sustainable, long-term programme. Sporadic, project-oriented, and short-term approaches are expensive and do not produce the clear results that contribute to lasting control of STH.
12. Strategically, the school community is an ideal focus for controlling STH and at the same time providing education on preventive behaviour. Its value will be undermined if programme development is not based on sound educational principles.

Other activities of YKB

YKB was established in 1980. At present other activities in its five clinics within and outside Jakarta consist of: maternal and child health and family planning services, including prevention of sexually transmitted infections and HIV/AIDS; annual physical examinations; laboratory tests of blood and urine; and mass chest X-ray using a mobile unit donated by the Chiba Health Service Association of Japan. Another laboratory is set up for quality assurance of intrauterine devices and condoms in collaboration with PATH (Program for Appropriate Technology in Health). YKB also undertakes research and training activities.

Conclusions

Learning from Japan's successful experiences in controlling intestinal worm infections, in 1987 YKB began to implement school parasite control with the collaboration of the Japan Association for Parasite Control and Japan Organization for International Cooperation in Family Planning .

A sustainable helminth control programme has been implemented in which educational activities have been integrated with the provision of regular stool examination and selective treatment. It has been able to sustain its activities over several years 12 years by using a "self-reliance" approach, charging beneficiaries of the programme small, affordable fees. With more support, the programme has the capacity to establish sustainable parasite control programmes in other parts of Indonesia.

Experience shows that the programme could develop beyond parasite control activities into a more comprehensive school health programme, benefiting not only pupils but also teachers. The synergistic effects of integrating the programme with existing YKB maternal and child health and family planning clinics have resulted in improved, sustainable, and affordable services that benefit the health and welfare of the people of Jakarta. Mass-coverage parasite control can thus serve to stimulate and improve health care in the community (Hara, 1985).

References

Hara T (1985) Promotion of mass parasite control programme as the trigger for community health development. In: *Notes for the APCO Integrated Project Training Course*. Bangkok, Faculty of Tropical Medicine, Mahidol University.

Kunii C (1992) *It all started from worms. The 45-year record of Japan's post-World War II national health and family planning movement.* Tokyo, Hoken Kaikan Foundation.

Margono, Sri S et al. (1999) Pengaruh Pengobatan Cacingan pada Anak Sekolah di Propinsi Lombok. [The effects of anthelminthic treatment on schoolchildren in Lombok Province.]

Ministry of Health (1995) *Laporan hasil survey prevalensi dan intensitas kecacingan pada anak sekolah di Sumatera Utara. [Report on the results of prevalence and intensity of infection of STH among primary schoolchildren in North Sumatra Province.]* Jakarta, Ministry of Health, Directorate General of Communicable Disease Control.

P4I (1992) *Menyambut harijadi P4I. [Celebrating the anniversary of P4I.]* Jakarta, Perkumpulan Pemberantasan Penyakit Parasitik (Indonesian Parasite Control Association).

Ratnawulan D (2000) *Laporan kegiatan laboratorium pemberantasan cacingan [Report on the activities of parasite control done by the central laboratory.]*

Sasongko A, Ratnawulan D, Lubis F (1999) *Laporan hasil pemberantasan cacingan di KEP, Seribu, Jakarta Utara. [Report on the result of STH control in Seribu Island, North Jakarta.]* Presented to the Vice-Governor of Jakarta Province.

Vajrasthira S, Waikagul J (1999) Laboratory diagnosis of soil-transmitted helminths. In: *Notes for the APCO Training Course for Senior Administrators and Leaders of Parasite Control Project for Community Health Development.* Bangkok, Faculty of Tropical Medicine, Mahidol University.

WHO (1991) *Basic laboratory methods in medical parasitology.* Geneva, World Health Organization, 1991.

The Seychelles experience in controlling helminth infections

Nicholas Shamlaye[1]
Ministry of Health
Victoria, Seychelles

Introduction

Intestinal parasite control programmes—aimed at controlling morbidity and, in the long term, reducing transmission of intestinal parasitic infections through different approaches—have been implemented in many developing countries (WHO, 1990). However, few have been planned at the national level with integration into existing health care systems and with efficient monitoring and information systems to assess the impact of control activities (Bundy et al., 1992).

In 1993, the Ministry of Health, in collaboration with the Ministry of Education and with technical support from the World Health Organization, developed the Seychelles Intestinal Parasites Control Programme (SIPCP); its goal was to reduce intestinal parasitic infections to a level at which they were no longer a public health problem. The Programme received financial support from the German Pharma Health Fund.

The Seychelles control programme was planned at the national level and from the start has been integrated into the existing primary health care system, thus reducing costs and human resource needs and strengthening the existing school health programme. The initiative has helped in promoting the collaboration of health staff at the central and district levels. Through the school system approach and with the support of the national press and media, the community has been mobilized and made aware of the need for prevention of intestinal parasitic infections and of other related health issues.

The implementation and management of the programme and analysis of the impact of 5 years of control activities are discussed in this paper.

The Seychelles archipelago

The islands of the Republic of Seychelles have a land area of 444 km^2 but are spread across 388 000 km^2 of the south-west Indian Ocean, between latitude 4°S and 10°S and longitude 46°E and 56°E. Altogether there are 115 islands: 42 are granitic and the remainder coralline.

[1] Dr N. Shamlaye, Environmental Health Section, Ministry of Health, P.O. Box 52, Victoria, Seychelles
Tel: +248 388191/388079, Fax: +248 224792

The Seychelles has an equable tropical climate influenced by monsoons. The warm and humid north-west monsoon blows from December to April and the south-east monsoon, which determines a cooler and drier season, from June to September. Average temperature ranges from 24°C to 31°C and rainfall averages 2200 mm annually. The country's economy is based on the major activities of tourism and fishing.

Mahé is the largest island of the Seychelles, covering an area of 152 km^2. Of the estimated total population of 81 000, some 90% live on Mahé and most of the remainder on the islands of Praslin and La Digue. Mahé is divided into five regions (North, South, East, West, and Central), which are made up of 25 districts in total. Victoria is the capital town and has a population of 12 000 (Ministry of Manpower, 1994).

Health facilities

The health system in the Seychelles provides essential health care services free of charge to every citizen regardless of financial means, thereby ensuring that no one is deprived of proper medical attention (Sullivan & Shamlaye, 1992).

The central hospital in Victoria, which has well-equipped clinical and public health laboratories, serves the North, Central and East regions of Mahé and is the main referral centre for specialized treatment. There are three other smaller hospitals—one in Anse Royale, Mahé, which is the referral centre for the South and West regions, and one each in Praslin and La Digue. Each centre has basic laboratory facilities.

There are 16 primary health care centres; personnel include school health nurses and environmental health officers (EHOs) who regularly visit all the district schools as part of their routine activities. All regional health centres have small laboratories for microscopic examinations of stool specimens.

Educational system

In Seychelles there are 39 primary and secondary schools spread throughout the country. The total enrolment is about 20 000 pupils.

A health-promoting school programme is currently being implemented. Its main activities include formal and informal curricula relating to healthy and safe school environments, provision of appropriate health services, and family and community participation. All activities are performed by the various health officials in collaboration with school management teams and teachers. A special programme of personal and social education is also carried out by specially trained teachers at primary and secondary levels.

Intestinal parasitic infections in the Seychelles

Intestinal parasitic infections were the main stimulus for the initiation of the Public Health Inspectorate in 1917. In 1925 efforts were made to tackle the hookworm problem in public latrines, with the result that the prevalence of hookworm infections was reduced from 90% to 41%. However, a lack of continuity together with misuse of anthelminthic drugs and poor community participation contributed to the persistence of intestinal parasitic infections.

A study performed in 1988 (Kitua, Shamlaye & Padayachy, 1989) indicated a prevalence rate of 84.4% for intestinal parasites, with a high proportion of multiple infections. Prevalence of *Trichuris trichuria* was 84%; figures for hookworm, *Ascaris lumbricoides*, *Giardia intestinalis*, and *Entamoeba histolytica* infections were 22%, 8%, 6%, and 5%, respectively (Bilo & Bilo-Groen, 1983).

Water and environmental sanitation

A treated water supply is available to 90% of the population. Nearly 89% of households have a flush water closet draining to septic tanks, soakaway pits, or field drain systems; 9% use pit latrines and aqua privies, leaving only 2.0% of households without proper toilet facilities.

Programme objectives and strategies

The general objective of the control programme (Ministry of Health, 1994[1]) was the reduction of intestinal parasitic infections to a level at which they were no longer a public health problem. The operational objectives were, within three years, to:

- — reduce the intensity of *A. lumbricoides* infections by 60% and of *T. trichiura* and hookworm infections by 30% in school-age children;
- — reduce the prevalence of *Strongyloides stercoralis* infection in the target population by 30%;
- — reduce the prevalence of amoebiasis in the target population by 40%;
- — make schoolchildren the main target group as they are more heavily infected, they suffer most from morbidity due to these infections, they are the most important source of infection, they can be easily reached in the schools, and health services can be delivered by teachers and health staff.

The main strategies that have been adopted in order to ensure a low-cost/high-benefit programme are periodic chemotherapy of the target population, health education and community participation, maintaining a high level of sanitation, and providing safe drinking-water supplies.

Programmes of this type require the collaboration and combined efforts of all concerned sectors. They need to be comprehensive and to cover basic aspects of sanitation such as construction and maintenance of latrines, prevention of conta-

[1] Ministry of Health, Seychelles, and Programme of Intestinal Parasitic Infections, Division of Communicable Diseases, World Health Organization. *A National Plan for the control of intestinal parasitic infections 1992–1995* (unpublished document, 1994).

mination of water supplies, and education of the population in matters of personal hygiene.

Management of the control programme

Integration of the control programme into the well-established primary health care system has benefited from being able to make full use of existing infrastructure and human and technical resources. The programme was implemented as a special programme under the Division of Disease Prevention and Control of the Ministry of Health. The programme activities are coordinated nationally by a Programme Manager with technical support from clinical and public health laboratories, the Health Education Unit, and the Epidemiology Unit. The school health nurses and EHOs, in close collaboration with the schoolteachers, are responsible for evaluation surveys, treatment campaigns, and health education activities in their particular health centre catchment areas.

At the start of the control programme in 1993, Ministry of Health staff were trained in this integrated approach. Special training courses in quantitative diagnostic techniques for stool examinination (Kato–Katz) and concentration techniques (formalin–ethyl acetate) (WHO, 1994) were organized for laboratory technicians from hospital and health centre laboratories. Nurses and EHOs were trained in prevention and control of intestinal parasitic infections. Discussions with physicians on diagnosis and treatment of intestinal parasitic infections were organized.

Since the programme is school-based, meetings were organized with Ministry of Education officials to explain the programme activities and to secure the active participation of teachers and school management bodies. With the Ministry's agreement, a training workshop on prevention and control of intestinal parasitic infections was organized for head teachers and social education teachers. The teachers' role as health educators was emphasized, as well as participation in motivating children during deworming campaigns and parasitological surveys.

The activities of EHOs in school settings have been stepped up, and other specific programmes—food control, water and sanitation, vector control, and health hazards control—have coordinated their work in improving the health standards of school premises and the personal hygiene behaviour of pupils and staff.

Treatment

Mebendazole (as 400-mg tablets) was chosen as the anthelminthic for treating schoolchildren (Albonico et al., 1994).

Mebendazole has been well accepted by the target population. Teachers and the general public noted the many immediate benefits of treatment and the fact that side-effects were negligible. The periodic deworming of schoolchildren has prompted the rest of the population to seek out treatment for themselves. In the past, deworming of schoolchildren during the school holidays was done routinely by many health-conscious parents, and the programme has now reactivated this practice. Mebendazole tablets have been made readily available by the Ministry of Health through the various health centres, and sales of deworming tablets in private pharmacies have increased.

This periodic chemotherapy of schoolchildren has also helped to promote health education at the district level and the integration of other programmes, such as sanitation control, at the level of the individual household. In the long-term, drug treatment will prevent healthy people becoming infected by reducing the sources of infection—contaminated soil, surface water, and general environment.

Community participation and health education

Print media (newspapers, posters, leaflets) and electronic media (radio, television, audiovisual aids) were extensively used for health education to increase public awareness of intestinal parasite control and to prompt people to take responsibility for control and preventive measures.

Since the start of the programme, preventive measures have been included as part of the national curriculum for both primary and secondary schools. Mobile health teams consisting of EHOs and school health nurses, in collaboration with social education teachers, have organized health education sessions and spread health messages to all classes of all schools in the country. In addition, a video on prevention and control of intestinal parasites, produced in the Seychelles, was widely distributed in all schools and health centres and was transmitted by the local television station.

Health education sessions on related subjects have been provided to other groups such as district parent–teacher associations, church congregations, staff at various workplaces, food handlers, and environmental field workers. Emphasis has been placed on increased contact with the population through regular health visits to households, especially among high-risk groups in the community.

Multisectoral collaboration

Collaboration between the various concerned sectors has been the key to success and has been actively developed and maintained through numerous meetings and contacts. Sanitation improvement and the provision of safe water supplies are being achieved through national efforts: the common goal is to ensure safe means for collection and disposal of human wastes and the availability of treated water to the whole population.

Results of parasitological surveys

In April–May 1993, faecal samples collected from 5% of all schoolchildren were examined by the Kato–Katz technique before the start of control activities. The same stool samples were also examined for protozoa and for *S. stercoralis* larvae by means of the formalin–ethyl acetate concentration technique. Quality control was undertaken at central level and involved the random examination of 10% of the slides. Sampling was randomized: 1075 children from primary, secondary, and preschool classes were chosen. Some children were interviewed, using a questionnaire, to assess their knowledge of intestinal parasites before intervention. Between June and July 1993, stool samples from more than 300 women attending antenatal clinics were also examined for helminths and protozoa; blood samples were taken for measurement of haemoglobin levels.

Data from the parasitological examination, recorded with name, age, sex, weight, and height of the subjects examined, were analysed using the EpiInfo package. The intensity of *Ascaris*, *Trichuris*, and hookworm infections was measured indirectly as egg counts: mean egg counts were calculated as arithmetic means and expressed as eggs per gram of faeces (epg).

The results from the parasitological survey of schoolchildren (n = 1058) and pregnant women (n = 338) are summarized in Tables 1 and 2. On average, about 60% were infected with one or more parasites, with significant variation by region. *Trichuris* was the most common parasite (prevalence of 53.3%), followed by *Ascaris* (17.7%). Hookworm infections were present in 6.3% of schoolchildren and in 8.6% of pregnant women, with low intensity. These results were judged to justify the intervention with mass treatment in schoolchildren without prior screening.

Only 14% of women had a haemoglobin level below 110 g/litre; in 3 cases (1%) the level was below 7 g/litre. No correlation could be found between intensity of hookworm infections and anaemia in the study population of pregnant women.

Results from the surveys in pregnant women are shown in Table 2. There was a significant reduction in prevalence (from 44.4% to 15.1%) and intensity (by 90%, 75%, and 73% of the baseline levels for hookworm, *Ascaris*, and *Trichuris* infection, respectively).

Comparing the results of the 1993 survey (baseline) and the 1998 survey (after 5 years of programme activities) revealed significant reductions in the prevalence and intensity of helminth infections (Table 1). Cumulative prevalence of intestinal parasites declined significantly, from 60.5% to 13%. The prevalence of *Ascaris lumbricoides* dropped by 79% from 17.7% to 0.9%, with a 95% fall in intensity of infection; *Trichuris* prevalence decreased by 60% from 53.3% to 5.3%, with a 92% lowering of infection intensity. For hookworm infection, prevalence dropped by 75% from 6.3% to 1.5% and intensity of infection fell by 50%.

Outcomes of the control programme

1. Increased health awareness among schoolchildren seems to have contributed to the lowering of parasite prevalence in general.
2. Increased health awareness has led to positive changes in personal habits, such

Table 1. Prevalence and intensity of intestinal parasites in schoolchildren of Seychelles, 1993–1994, 1996, and 1998

Parasite	1993		1994		1996		1998		% reduction 1993/1998	
	Prevalence (%)	Intensity (epg)	Prevalence (%)	Intensity (epg)	Prevalence (%)	Intensity (epg)	Prevalence (%)	Intensity (epg)	Prevalence (%)	Intensity (%)
Ascaris	17.7	1617	4.4	244	3.7	134	0.9	200	79	95
Trichuris	53.3	782	27.3	367	21.5	299	5.3	48	60	92
Hookworm	6.3	40	4.2	27	1.6	7	1.5	20	75	50
Strongyloides	1.1	—	0.3	—	0.4	—	—	—	100	—
Entamoeba histolytica	4.6	—	1.1	—	2.8	—	2.5	—	39.1	—
Giardia	3.3	—	2.6	—	2.4	—	2.7	—	18.2	—
Cumulative prevalence	**60.5**	—	**33.8**	—	**28.6**	—	**13.0**	—	**78.5**	—

Table 2. Prevalence and intensity of intestinal parasites in pregnant women at baseline (1993) and after two years (1995) of programme activities

Parasite	1993 Survey		1995 Survey		% Reduction 1993/1995	
	Prevalence (%)	Intensity (epg)	Prevalence (%)	Intensity (epg)	Prevalence (%)	Intensity (epg)
Ascaris	9.8	641	3.2	163	67	75
Trichuris	36.1	302	9.9	82	73	73
Hookworm	8.6	42	3.0	4	65	90
Stronglyloides	2.7	—	1.5	—	NS[a]	—
Entamoeba histolytica	5.3	—	4.2	—	NS[a]	—
Giardia	0.6	—	2.2	—	NS[a]	—
Cumulative prevalence	**44.4**	—	**15.1**	—	**70**	—

[a] NS = not significant.

as frequent washing of hands, wearing of shoes, washing of fruits and vegetables.

3. Children have acted as health messengers to their parents and their households by transmitting whatever they have learned at school.
4. The many activities have consolidated the health team concept of contributing not only to primary health care but also to curative health services.
5. There has been a strengthening of intersectoral collaboration between the health sector and other key development areas—education, environment, communication, public utilities, agriculture, planning, housing.
6. The active participation and acceptance by the community is well appreciated and is essential for the future sustainability of the control programme.
7. Radio phone-in programmes—people call in and ask for clarifications and advice.
8. Mebendazole has been made readily and cheaply available to the general population at the various health centres.
9. The school teaching curriculum now includes the subject of intestinal parasites and their control and prevention.

Conclusion

The success of the control programme has been due to strong political commitment through the support and action of the health and education ministries and to effective intersectoral collaboration between the various authorities concerned. Public awareness of intestinal parasite infections, and of the need to control them, has been raised in an effort to change health behaviours. The efficient and supportive school system that exists in the Seychelles, within which the health-promoting school concept is being effectively implemented, has created the conditions necessary for the smooth running of the programme's activities. The long-term objective is to consolidate and maintain the school-based interventions for prevention and control of parasite infections and to expand the efforts to the general community. The programme's impact in reducing parasitic infections in the untreated adult population through a reduction in transmission will in future be evaluated using records from the clinical, public health, and regional diagnostic laboratories. During implementation of the programme in schools, contact between schoolteachers, school health nurses, EHOs, and parents has increased. The phone-in activity on the local radio proved particularly effective. As a result of the regular education talks and related activities, parents readily accepted the activities in school and in the wider community.

EHOs are actively engaged in working with rural communities at higher risk of infection to improve their sanitation and water supplies. To promote preventive measures, district EHOs and the health team carry out regular visits to individual households.

For reasons of long-term cost-effectiveness, the programme activities are gradually being integrated into the school health system rather than being carried out in isolation.

The Seychelles Intestinal Parasite Control Programme is an example of effective control of a public health problem through existing health care facilities. The Programme has facilitated collaboration between health, education, and other ministries for the control of communicable diseases. It can be seen as a model for other developing countries: the comprehensive approach can provide an opportunity to strengthen an existing primary health care programme and become an entry point for controlling other diseases. The results are promising and represent an improvement in the provision of sustainable and affordable services that benefit the health and welfare of the people of Seychelles.

Acknowledgement

The author is most grateful to the personnel of the Ministry of Health, particularly the laboratory technicians, school health nurses, EHOs, and district managers, and to the collaborators in other ministries, especially the schoolteachers and school management teams.

The financial support to the SIPCP provided by the German Pharma Health Foundation, through the World Health Organization, is gratefully acknowledged.

Special thanks are also extended to the children and adults of Seychelles.

References

Albonico M et al. (1994) A randomised controlled trial comparing mebendazole 500 mg and albendazole 400 mg against *Ascaris, Trichuris* and the hookworms. *Transactions of the Royal Society of Tropical Medicine and Hygiene*, 88:585–589.

Bilo HJ, Bilo-Groen CE (1983) Worm, giardia and amoebic infestations on Praslin, Seychelles. *Tropical and Geographical Medicine*, 35:179–180.

Bundy DAP et al. (1992) Evaluating measures to control intestinal parasitic infections. *World Health Statistics Quarterly*, 45:168–179.

Kitua A, Shamlaye C, Padayachy J (1989) *The prevalence of intestinal parasitic infections in Seychelles—1988*. Seychelles, Health Statistics Section, Epidemiology and Research Division, Ministry of Health.

Ministry of Manpower (1994) *Population vital statistics, 1994*. Seychelles, Statistical Bulletin, Management and Information Systems Division, Ministry of Manpower.

Sullivan FM, Shamlaye C (1992) Primary medical care in Seychelles. *Tropical Doctor*, 22:100–104.

WHO (1990) *Informal Consultation on Intestinal Helminth Infections, Geneva, 9–12 July 1990*. Geneva, World Health Organization (document WHO/CDS/IPI/90.1).

WHO (1994) *Bench aids for the diagnosis of intestinal parasites*. Geneva, World Health Organization.